Experiencing Geometry, Physics, and Biology

Edition Angewandte – Book series of
the University of Applied Arts Vienna

Universität für angewandte Kunst Wien
University of Applied Arts Vienna

Georg Glaeser • Franz Gruber

Experiencing Geometry, Physics, and Biology

DE GRUYTER

Georg Glaeser
Head of the Institute of Arts & Technology / Geometry,
University of Applied Arts Vienna, Austria

Franz Gruber
Institute of Arts & Technology / Geometry,
University of Applied Arts Vienna, Austria

Project Management "Edition Angewandte" on behalf of the
University of Applied Arts Vienna:
        Anja Seipenbusch-Hufschmied, A-Vienna
Content and Production Editor on behalf of the Publisher:
        Katharina Holas, A-Vienna

Proofreading/Copyediting:
        Eugenie Maria Theuer, Irene Karrer, Julia Weber, A-Vienna
Translation from German into English:
        Eugenie Maria Theuer, Irene Karrer, A-Vienna
Layout, cover design, and typography:
        Georg Glaeser, A-Vienna
Printing: Beltz Grafische Betriebe GmbH, D-Bad Langensalza

Library of Congress Control Number: 2023946837

Bibliographic information published by the German National Library
The German National Library lists this publication in the Deutsche
Nationalbibliografie; detailed bibliographic data are available on the
Internet at http://dnb.dnb.de.

ISSN 1866-248X
ISBN 978-3-11-136523-7
e-ISBN (PDF) 978-3-11-136578-7

Ursprünglich veröffentlicht in deutscher Sprache:
Geometrie, Physik und Biologie erleben von Prof. Dr. Georg Glaeser
© Springer-Verlag GmbH, DE, ein Teil von Springer Nature, 2020.
Alle Rechte vorbehalten.

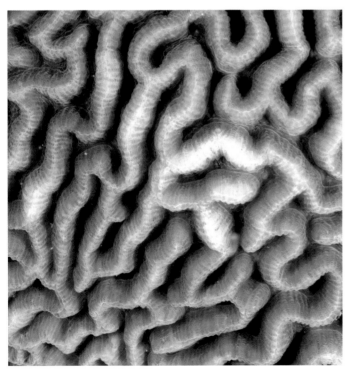

Above: Hilbert curve on a sphere
Below: Photograph of a brain coral

Preface

How the software was created

Geometry geometry geometry try geometry geometry geometry The two authors have spent many years working together at the Department of Geometry of the University of Applied Arts Vienna. Over the years, this Department saw the emergence of a heterogeneous, and perhaps for this exact reason, successful team dedicated to the development of geometry softwares.

It started with the CC++ programming environment Open Geometry, which was developed by Georg Glaeser in collaboration with members of the Vienna University of Technology (Hellmuth Stachel) and the University of Innsbruck (Hans-Peter Schröcker). We are indebted to Peter Calvache for his decisive impulses, especially in the development of a professional interface, as well as to Günter Wallner and, more recently, Christian Clemenz for their valuable contributions. Many of the integrated simulations were created by Christian. We have also received remarkable input by several users of Open Geometry, as well as by students of the University of Technology Vienna and the University of Applied Arts Vienna. Boris Odehnal, meanwhile Professor of Geometry at the University of Applied Arts Vienna, has kindly provided steady support to the authors when tackling difficult geometrical questions.

Complex questions and realistic simulations

Franz Gruber developed a remarkable skill for programming with Open Geometry and was able to solve more complex problems using realistic simulations due to his profound knowledge of physics. His unexpected death in September 2019 was a great loss for the team. This tragic event, however, strengthened the desire to make the software, which was significantly influenced by him, available to the public.

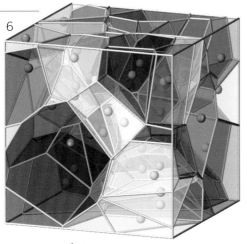

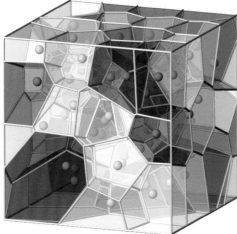

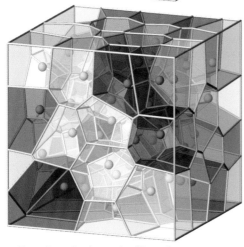

Foam formation in a cube (Voronoi diagram). Cell optimisation.

The software along is not enough

At some point in the process, we realised that merely "making the software available online" (the software had by then received the name Cross-Science) would not achieve its purpose: its themes are too specific and sometimes also too complex to be shared without further explanation. So, a "companion book" was needed.

Not just a description of the software!

This book has been created as a companion book for the freely available software Cross-Science. The software enables you to use and navigate around 140 individual, independent, interactive applications from the fields of biology, geometry, and physics. Cross-Science is in the public domain, and everyone who is interested is invited to "play around with it". However, this book is not meant to be merely a manual of the software package: nowadays many people, especially younger generations, can use softwares without reading any manuals. Yet, even these people will lack a deeper understanding of the majority of animations included in the software.

Embedding of videos

During the creation of this book, we became increasingly committed to the idea of making the results of the software directly available in the form of videos that can be accessed via QR codes. So, while leafing through the pages, readers can watch the accompanying animations or short films without having to rely on any particular hardware support. We have also added videos and animations that were not generated by computers but were created instead with other software systems: slow motion, time-lapse videos (e. g. of flying insects, dripping water faucets or slowly revolving clockworks), or the beautiful animations of our talented temporary coworker Meda Retagan. At the end of the book, you will find a list of all cited video sequences including their authors.

Who is this book for?

This book has been written for people that have a general affinity for the sciences and also for technology. This group of people could include lecturers and students but also younger pupils with an interest in these areas.

Demo video
http://tethys.uni-ak.ac.at/cross-science/3d-foam.mp4

What will you find in this book?

This book has been divided into fifteen chapters, which focus on different programmes dealing with themes that are similar to a certain extent – for instance, kinematics, single-curved and double-curved surfaces, biological mechanisms, photography, or fractals. For practical purposes, the chapters have been divided into double pages, which can be read in any order. Usually, each double page thematically covers one of the 140 programmes but they are not limited to this.

Let yourself be surprised by details!

If you have prior knowledge of geometry, physics, and biology, you might still learn about hitherto neglected areas after reading one double page or the other. It is almost guaranteed that even readers who are well familiar with the material will gain and enjoy new insights while reading this book. If you stumble upon a topic in which you are not particularly interested at the moment, then you can simply skip a double page without losing track.

Further reading "in the same style"

This book often discusses themes that have also been explored in earlier books – sometimes from a different perspective. The following books have also partially been illustrated by means of the software that is described here. Here is the list of books where you can learn more about examples that are discussed in the present book (which also re-uses illustrations from the earlier books):

[1] G. Glaeser: *Geometry and its Applications in Arts, Nature and Technology*. 2nd edition, Springer, Vienna / New York, 2020.
[2] G. Glaeser: *Nature and Numbers: A Mathematical Photo Shooting*. Edition Angewandte, Vienna, 2014.
[3] G. Glaeser, H. F. Paulus: *The Evolution of the Eye*. Springer Nature, Berlin / Heidelberg, 2014.
[4] G. Glaeser, H. F. Paulus, W. Nachtigall: *The Evolution of Flight*. Springer Nature, Berlin / Heidelberg, 2017.
[5] G. Glaeser, W. Nachtigall: *The Evolution and Function of Biological Macrostructures*. Springer Nature, Berlin / Heidelberg, 2019.
[6] G. Glaeser: *Moonstruck: The Interplay of Celestial Bodies in Pictures*. De Gruyter, Edition Angewandte, Berlin / Boston, 2021.

A somewhat clunky looking "source material" turns – through constant shrinking of the surface – into an organic looking and simultaneously optimal triangulated structure.

Demo video
http://tethys.uni-ak.ac.at/cross-science/reducing-the-surface.mp4

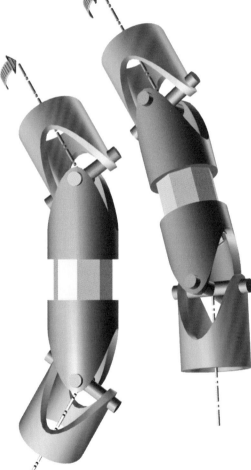

The Cardan joint has important applications in technology. It transfers a rotation from one axis to an intersecting axis. It can only show its true power though when at least two of these joints are combined.

Demo video
http://tethys.uni-ak.ac.at/cross-science/cardan-joints2.mp4

Table of Contents

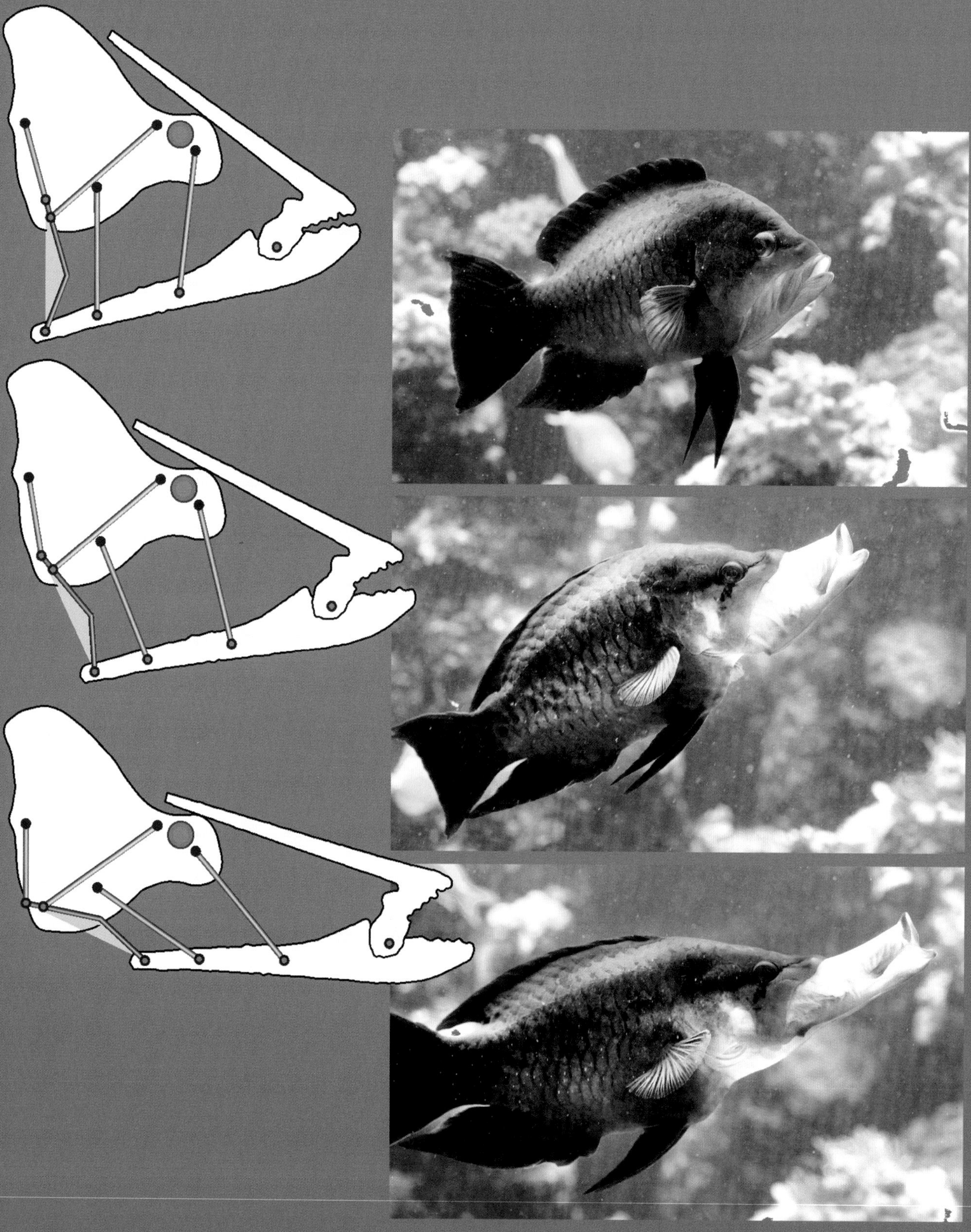

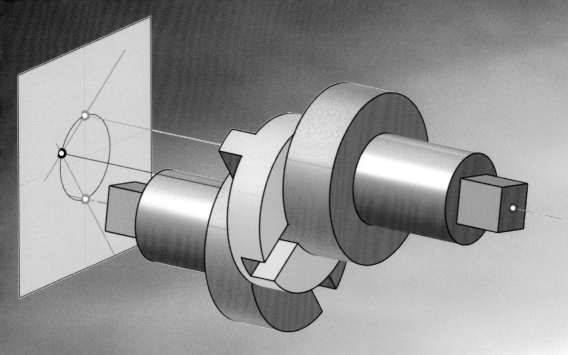

Kinematics:
Motions in Nature and Technology

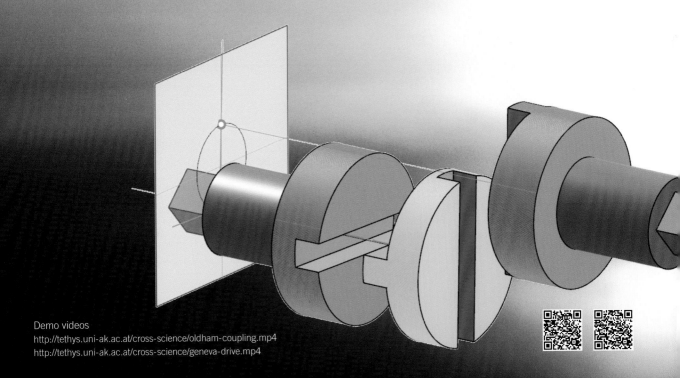

Demo videos
http://tethys.uni-ak.ac.at/cross-science/oldham-coupling.mp4
http://tethys.uni-ak.ac.at/cross-science/geneva-drive.mp4

Pliers and Jaws

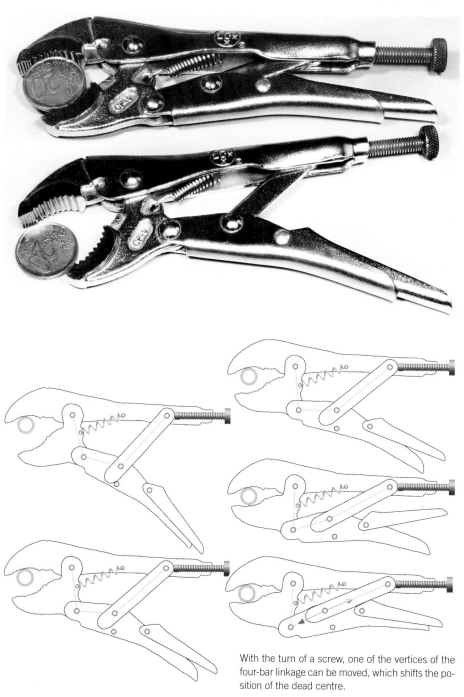

A practical securing mechanism

A plier is a manual tool with which objects can be held. Locking pliers can be locked into position. A lever is used to release them from their locked position.

Taking advantage of the "dead centre"

Due to the mechanism described above, this type of pliers can exert a great amount of force.

Locking pliers can be opened with a lever that overcomes the dead centre and thus releases the work piece. Applications include, for instance, the loosening of very tight screws, the securing of work pieces, and using the pliers as a "third hand".

Adapting to the object

The image series below shows how the locking position can be adapted to the diametre of the object to be locked. A screw is used for this purpose. Maximum forces are at play in the locked position. If the screw is correctly adjusted, the locking process will require hardly any exertion, and the force that holds the work piece is enormous.

Pliers in the animal kingdom

On the next page, we will explore a mechanism that is not unlike the one described on this page. It enables monitor lizards to flex their plier-shaped muzzles upward and downward.

With the turn of a screw, one of the vertices of the four-bar linkage can be moved, which shifts the position of the dead centre.

Demo video
http://tethys.uni-ak.ac.at/cross-science/wrench.mp4

Four-joint chain inside a monitor lizard's head

As the sketch here (Werner Nachtigall) shows, a kinematic chain consists of four elements 1 to 4 which are flexibly connected to each other. The system thus also features four joints A to D. The movement of such chains is "constrained". For instance, imagine that you are holding element 4 and rotating element 1 around its joint A: then element 3 will move back and forth in a defined manner, because this movement is transferred from element 1 to 3 by element 2, which is linked in between.

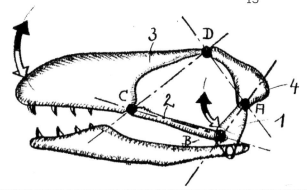

Simultaneous opening

When the lower jaw is lowered, muscles in the monitor's head will rotate bone 1 inside joint A to the top front. As a consequence, the upper jaw is automatically lifted, and vice versa. The advantages lie in how the two jaws can move against each other like scissors, as shown by the animation below.

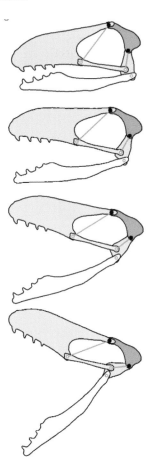

Komodowaran
(*Varanus komodoensis*)

Demo videos
http://tethys.uni-ak.ac.at/cross-science/iguana-bite.mp4

...-ak...
...org.glaeser@uni-ak.ac.at

Cracking Nuts and Devouring Prey

Grain eating with pincer principle

Grain-eating parrots (pictured here, a cockatoo) move the lower and upper parts of their beaks against each other. When the lower jaw is lowered, the upper jaw is lifted – and vice versa.

This is achieved by a forced coupling of the two beak halves, which consists of bone elements inside the head. Such a forced coupling is beneficial during grain eating. If one of the beak halves were fixed and the other beak half pressed against it, then the grains could easily slip out. Evolution has generated a more efficient principle: the two beak halves move, like the gripping jaws of pincers, against each other.

Beak is also used for climbing

Such "pincer beaks" also facilitate climbing. Parrots are known to use their beaks like a third limb, holding onto twigs or ledges while they look for support with their feet.

It can get more complicated

On this page we will introduce two mechanisms that are "implemented" in two predatory fish of different sizes: the harmless looking and relatively small sling-jaw wrasse *Epibulus insidiator* (p. 10) and the quirky looking goblin shark *Mitsukurina owstoni*, which usually lives at greather depths.

Jaw unfolding at the speed of light

Sling-jaw wrasses have developed a sneaky trick for hunting in coral reefs: While preying, they unfold their jaws rapidly into a long tube with which they suck in small fish. Intermediate positions of this motion have been captured in the animation in the left-hand column. This attack catches prey fish by surprise because fish usually measure the potential danger of predatory fish by their size, speed, and distance to them.

Protruding jaw

Goblin sharks have long protruding noses that remain almost immobile while the sharks grab prey. First the lower jaw snaps down, and then the two jaws are catapulted forward. The underlying kinematic motion is three-dimensional and is not easy to comprehend. In the intermediate positions, we can see that the points marked red remain fixed. Both jaws are divided into two parts, and the left and right part are elastically connected to each other.

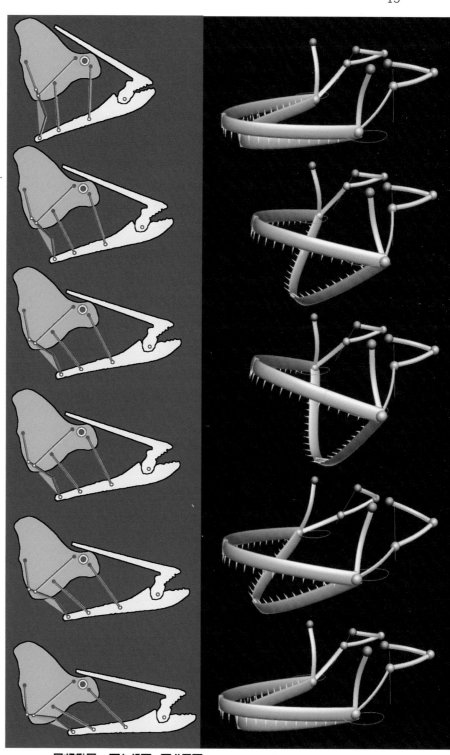

Demo videos
http://tethys.uni-ak.ac.at/cross-science/jaws.mp4
http://tethys.uni-ak.ac.at/cross-science/goblin-shark.mp4
https://www.youtube.com/watch?v=P8TT90LWYaw

Converting Translation into Rotation

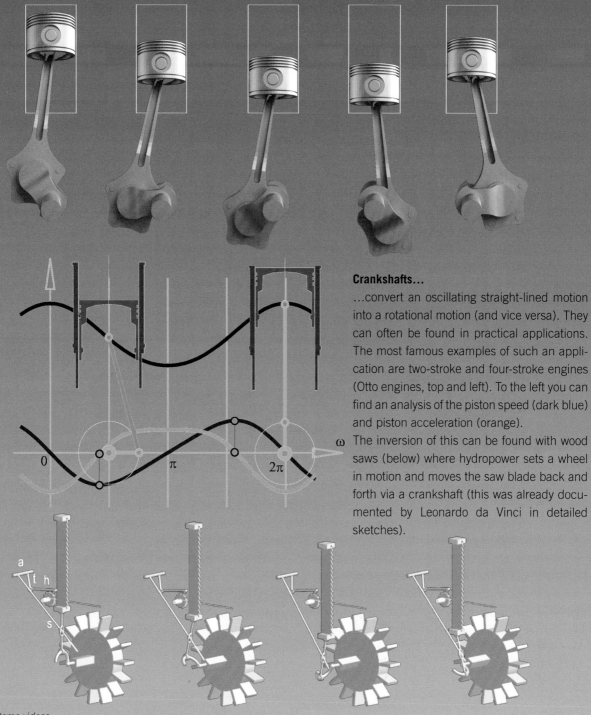

Crankshafts...

...convert an oscillating straight-lined motion into a rotational motion (and vice versa). They can often be found in practical applications. The most famous examples of such an application are two-stroke and four-stroke engines (Otto engines, top and left). To the left you can find an analysis of the piston speed (dark blue) and piston acceleration (orange).

The inversion of this can be found with wood saws (below) where hydropower sets a wheel in motion and moves the saw blade back and forth via a crankshaft (this was already documented by Leonardo da Vinci in detailed sketches).

Demo videos
http://tethys.uni-ak.ac.at/cross-science/piston-otto-engine.mp4
http://tethys.uni-ak.ac.at/cross-science/otto-engine-analysis.mp4
http://tethys.uni-ak.ac.at/cross-science/leonardos-saw.mp4

The radial engine

A remarkable generalisation of crankshafts can be found in the engines of propeller airplanes. These engines have several pistons driving the propeller shaft (marked in red). The corresponding video helps in providing a better understanding of this motion.

Demo video
http://tethys.uni-ak.ac.at/cross-science/radial-engine.mp4

Steam Locomotive

Until the middle of the 20th century, rail transport was dominated by steam locomotives. We will discuss two aspects using the Liliput locomotive in Vienna's Prater (built in 1928, see video) as a model: propulsion and gear shifting. The numbers 1 to 5 refer to the images on the right-hand page.

The piston steam engine

The burning of charcoal generates water steam that is pressed into the control cylinder from above (vertical red inlet). There the steam is released alternatingly through symmetrical valves. This creates a difference in pressure that causes a slider to move back and forth. Through corresponding slit openings, the actual piston then moves inside the steam cylinder underneath – a motion that is also created by pressure difference. The movement of the piston is now converted into rotation energy to drive one of the wheels.

The Walschaerts valve gear

This non-trivial mechanism was developed – apparently independently from each other – in Germany and Belgium in the 1840s, and it represented a masterstroke in engineering back then. Even with an animation video, it takes some time to understand the details of the multi-unit gearbox. It is remarkable that, in addition to the neutral position (image 1), the machine has respectively two forward gears (images 2 and 3) and two reverse gears (now image 4).

The back and forth motion of the piston is converted into the rotation of a single wheel (large wheel in the middle). Via a hinge parallelogram (marked red in image 5), this wheel drives the other two wheels of the locomotive.

Demo video and additional information
http://tethys.uni-ak.ac.at/cross-science/liliput-train.mp4
https://de.wikipedia.org/wiki/Dampflokomotive

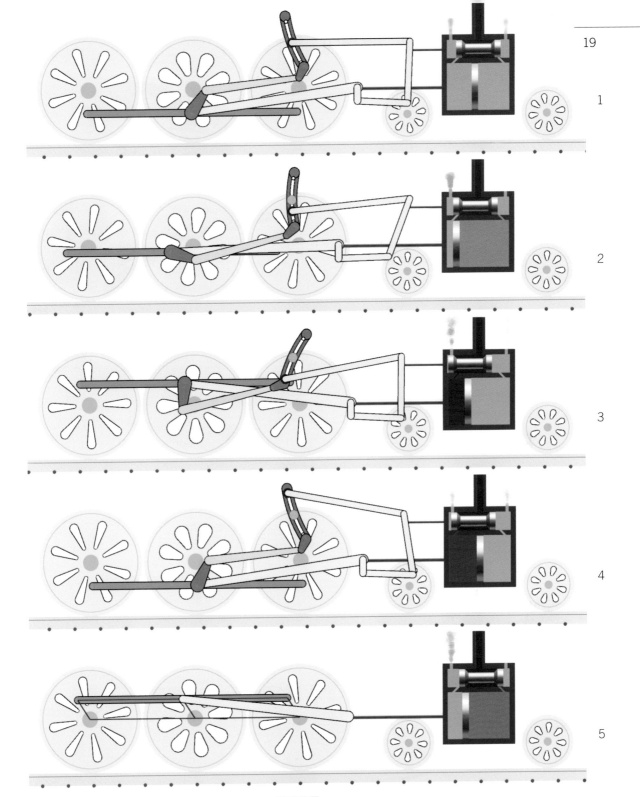

1

2

3

4

5

Demo video
http://tethys.uni-ak.ac.at/cross-science/lokomotive.mp4

Cardan Shafts

Transmission of torque

Cardan shafts enable a robust torque transmission in a kinked shaft train (the rotational axes must intersect each other). The kink angle can vary in the process (images to the right vs. images below), though only up to a limited angle of about $\pm 45°$.

"Cardan errors"

In the simplest case (single Cardan joint), the angular velocity of the (yellow) output shaft does not equal the (blue) drive velocity and oscillates periodically. The deviation increases with the kink angle, but the shaft becomes blocked at a kink angle greater than als $45°$ (see first video).

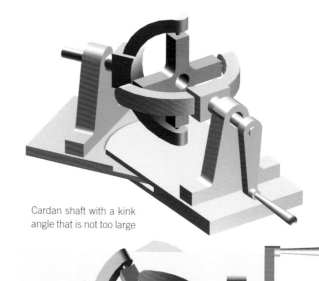

Cardan shaft with a kink angle that is not too large

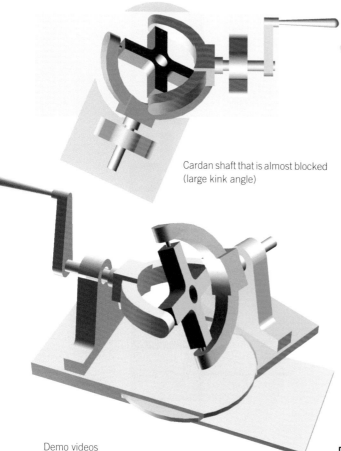

Cardan shaft that is almost blocked (large kink angle)

Two consecutive shafts

The flaw of a non-constant angular velocity can be remedied by – systematically – "linking together" two shafts. We insert an additional auxiliary axis (orange connecting piece), which joins the two axes in an angle-bisecting direction: the auxiliary axis intersects the two shafts at the same distance from the intersection point of the two axes. The Cardan joints' distance from each other can be arbitrary within certain limits, because it can be bridged with the connecting piece (see image at the top on the right-hand page as well as the middle-right image or second video).

Demo videos
http://tethys.uni-ak.ac.at/cross-science/cardan-joint.mp4
http://tethys.uni-ak.ac.at/cross-science/two-cardan-joints.mp4

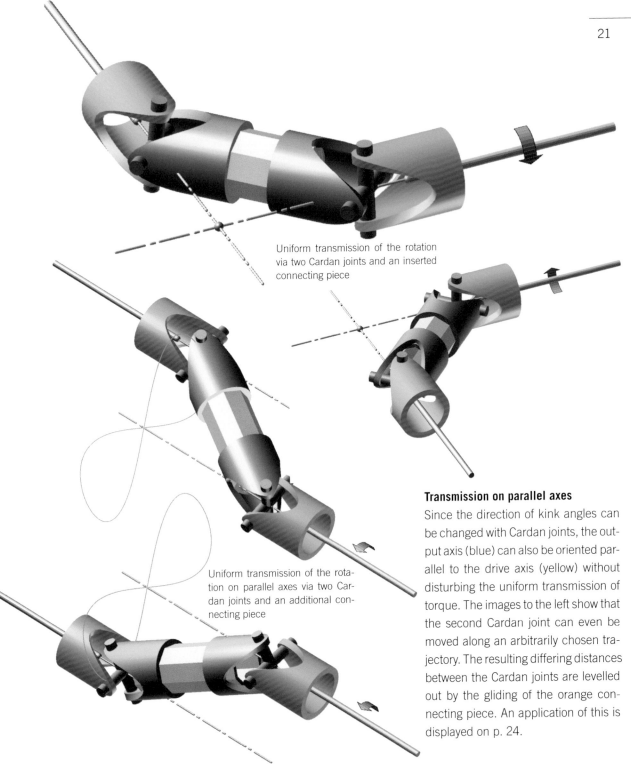

Uniform transmission of the rotation via two Cardan joints and an inserted connecting piece

Uniform transmission of the rotation on parallel axes via two Cardan joints and an additional connecting piece

Transmission on parallel axes

Since the direction of kink angles can be changed with Cardan joints, the output axis (blue) can also be oriented parallel to the drive axis (yellow) without disturbing the uniform transmission of torque. The images to the left show that the second Cardan joint can even be moved along an arbitrarily chosen trajectory. The resulting differing distances between the Cardan joints are levelled out by the gliding of the orange connecting piece. An application of this is displayed on p. 24.

Demo videos
http://tethys.uni-ak.ac.at/cross-science/cardan-parallel.mp4

Wide-Angle Cardan Shafts

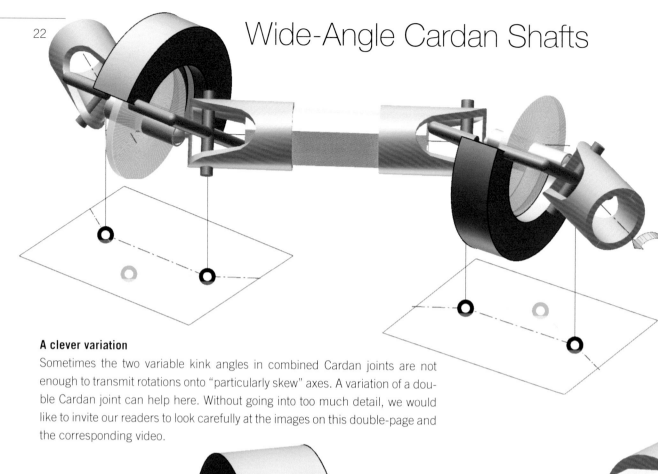

A clever variation

Sometimes the two variable kink angles in combined Cardan joints are not enough to transmit rotations onto "particularly skew" axes. A variation of a double Cardan joint can help here. Without going into too much detail, we would like to invite our readers to look carefully at the images on this double-page and the corresponding video.

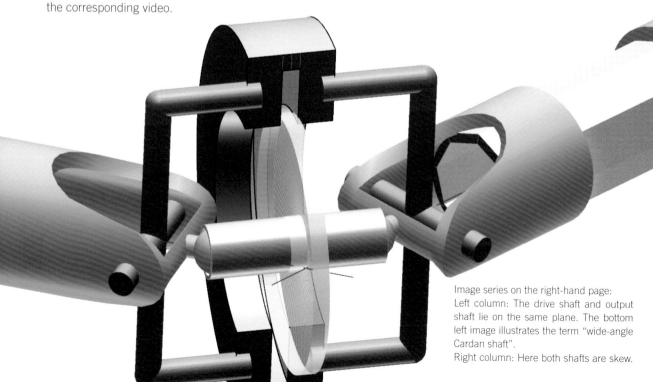

Image series on the right-hand page:
Left column: The drive shaft and output shaft lie on the same plane. The bottom left image illustrates the term "wide-angle Cardan shaft".
Right column: Here both shafts are skew.

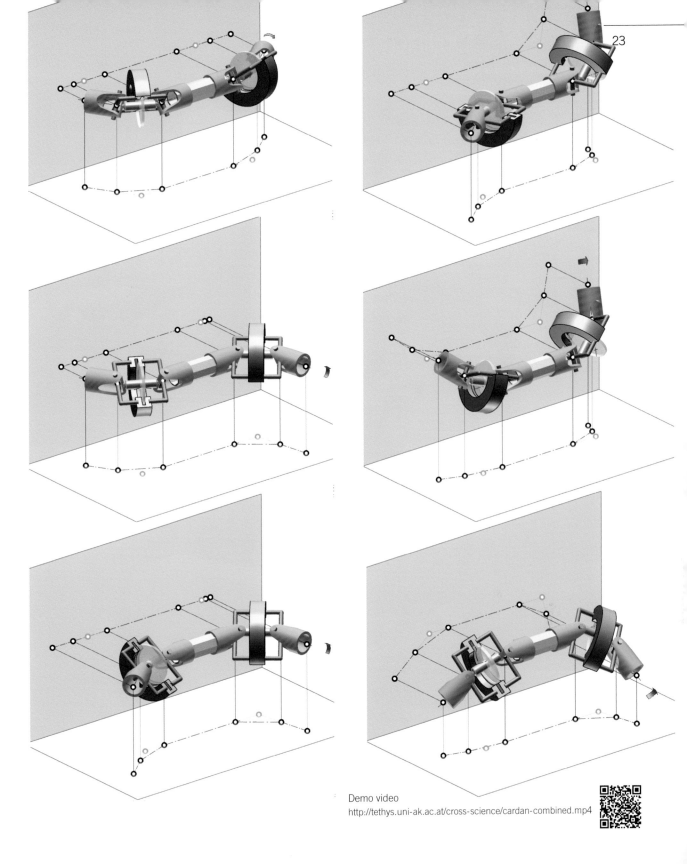

23

Demo video
http://tethys.uni-ak.ac.at/cross-science/cardan-combined.mp4

Drilling Square Holes

Rotating an orbiform inside a square

An equilateral triangle with side length a can be transformed into a so-called "orbiform" by placing a compass point at the vertices and drawing arcs through the other two vertices. This type of triangle can be arbitrarily rotated inside a square with side length a.

Never reaching the vertices of the square

During rotation the square's vertices are not milled out. The resulting fillets may look circular, but they are actually parts of ellipses (see next page).

A trick using a template

During rotation the centre of the orbiform will necessarily move in an oval shape. While drilling, however, you want to transform the rotation of a wave (a drill!) into a uniform rotation of the oval. This can be achieved through a drilling template and by means of two Cardan joints.

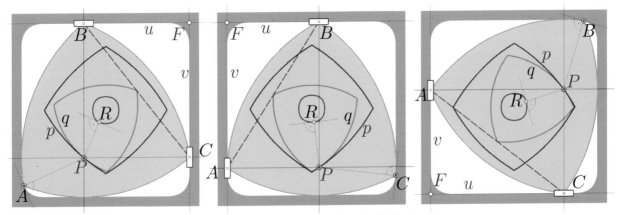

How exactly does the orbiform move?

The sketch above shows: The orbiform (named Reuleaux triangle after its inventor) touches the square with its boundary circles at exactly two points while two of its vertices always move along a straight line. This leads to a contradiction because the two points are als the centres of the circles.

A series of elliptical motions

Since the distance between two vertices is always constant, we always get an elliptical motion where a rod of constant length is guided with its two end points along two straight (perpendicular) lines. If the guiding lines change, then so does the elliptical motion.

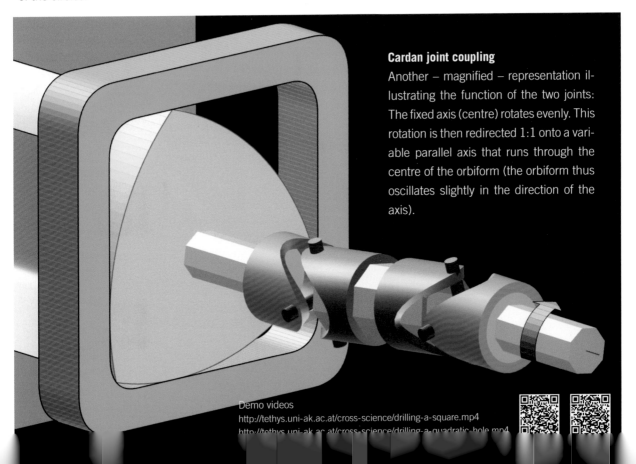

Cardan joint coupling

Another – magnified – representation illustrating the function of the two joints: The fixed axis (centre) rotates evenly. This rotation is then redirected 1:1 onto a variable parallel axis that runs through the centre of the orbiform (the orbiform thus oscillates slightly in the direction of the axis).

Demo videos
http://tethys.uni-ak.ac.at/cross-science/drilling-a-square.mp4
http://tethys.uni-ak.ac.at/cross-science/drilling-a-quadratic-hole.mp4

R6 Mechanisms and Kaleidocycles

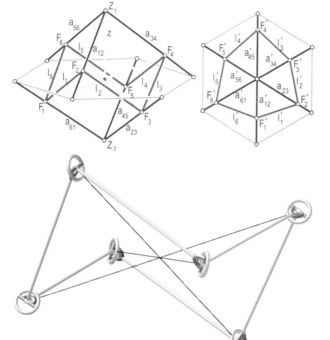

Escher kaleidocycles

The following only works under very specific circumstances: Take the net of a polyhedron, assemble it correctly, and then move in thousands of different ways.

Six rotations (one "R6 mechanism")

The construction of the famous kaleidocycle by Maurits Cornelis Escher is based on the so-called R6 mechanism, where a closed polygon is formed by two pairs of three intersecting rods that are symmetrically arranged around a common rotation axis. The polygon's edges are the common normals of each rod pair. The name of the mechanism is derived from the six rotations that are involved in it.

The motion is "spatially constrained"

The construct moves in a precisely defined manner depending on which points of the mechanism are held fixed.

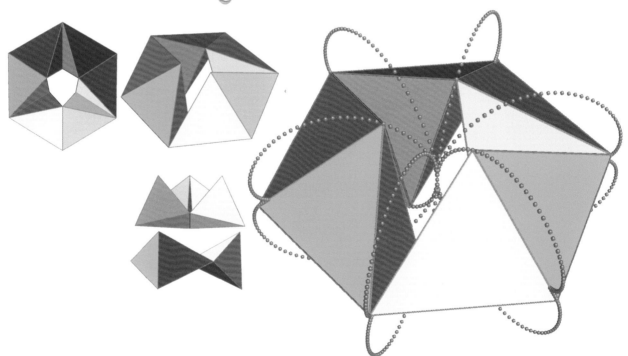

Demo videos
http://tethys.uni-ak.ac.at/cross-science/r6-mechanism.mp4
http://tethys.uni-ak.ac.at/cross-science/r6-mechanism2.mp4
http://tethys.uni-ak.ac.at/cross-science/caleidocycle1.mp4

Tetrahedron chains

A closer look reveals that the kaleidocycles are formed from special sets of congruent tetrahedrons.

If a kaleidoscope is turned around in such a way that the trajectory of the vertices aligns with the meridian layer, then you will get a result similar to that in the images at the bottom of the left page or on the right of this page.

Fixing the edges?

The blue series on the right illustrates that one can also fix individual tetrahedron edges. This causes the tetrahedron chain to move differently relatively to the space – the intermediate shapes will occur at different positions in the space.

Intermediate shapes

If we choose an edge length for the congruent building blocks as in the example of the blue chain, then we will get some remarkable intermediate shapes. In this concrete case, the chain closes twice, and there is one position where it fits into a cube. This shape discovered by Paul Schatz is known as "cube belt".

Demo videos and websites
http://tethys.uni-ak.ac.at/cross-science/inverse-r6.mp4
http://tethys.uni-ak.ac.at/cross-science/caleidocycle2.mp4
http://www.fzk.at/wuerfel/schatz.html

Rolling and Turning: The Oloid

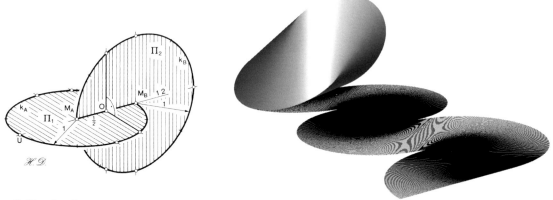

Rolling two beer mats

The following experiment works very well: Take two circular beer mats of equal size, cut a slit into one of them, and connect the two discs through the slit – vertically – so that the centre of each disc lies on the circular edge of the other (left image at the top). This construct can easily be rolled across a table top, where it would move in a basically straight line but still "wobble" in a funny way.

This hull is a developable and known as oloid

In any position the two circles are touching the table top at one point each. The connecting straight line between the two points lies entirely on the plane of the table surface.

The entirety of all connecting straight lines forms a developable (simple curved) surface with two sharp circular edges. One can let a "moving trihedron" travel across the surface (first image series below), with one axis always running through each point of a generating surface normal and the two other axes spanning the tangential plane of the surface along the generatrix.

All generatrix trajectories have the same length

It can be proven that the trajectory from one circle to the other is of constant length, which enables us to fix this trajectory and let the oloid "wobble" below it (second image series below).

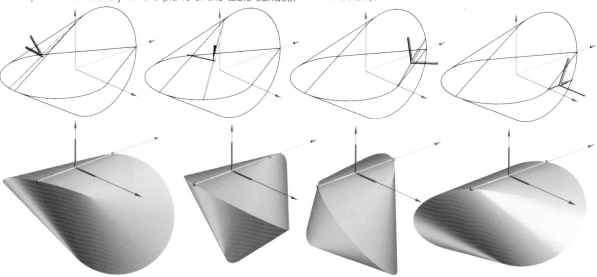

Demo videos and literature
https://www.heldermann-verlag.de/jgg/jgg01_05/jgg0113.pdf
http://tethys.uni-ak.ac.at/cross-science/rolling-oloid.mp4

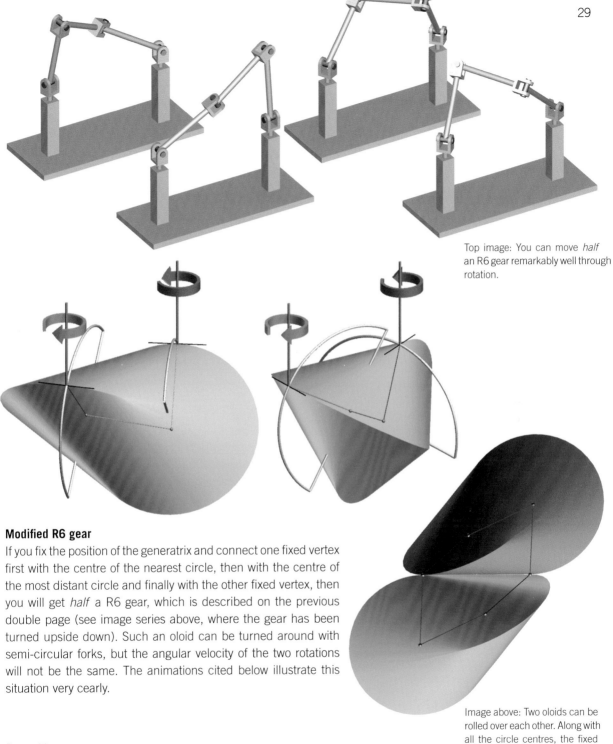

Top image: You can move *half* an R6 gear remarkably well through rotation.

Modified R6 gear

If you fix the position of the generatrix and connect one fixed vertex first with the centre of the nearest circle, then with the centre of the most distant circle and finally with the other fixed vertex, then you will get *half* a R6 gear, which is described on the previous double page (see image series above, where the gear has been turned upside down). Such an oloid can be turned around with semi-circular forks, but the angular velocity of the two rotations will not be the same. The animations cited below illustrate this situation very cearly.

Image above: Two oloids can be rolled over each other. Along with all the circle centres, the fixed points then form an R6 gear.

Demo videos
http://tethys.uni-ak.ac.at/cross-science/oloid-motion.mp4
http://tethys.uni-ak.ac.at/cross-science/oloid.mp4

Gearwheels:
Precise and Robust

Another Proven Deflection Method

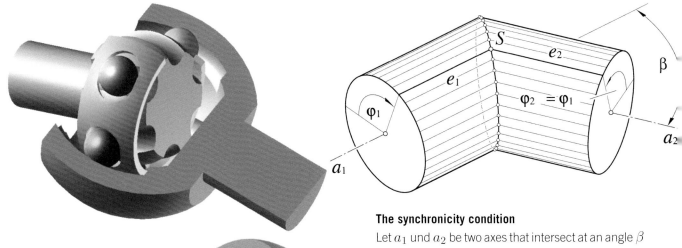

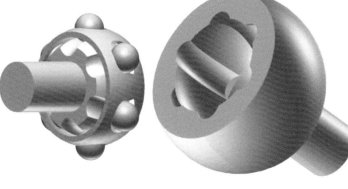

The synchronicity condition

Let a_1 und a_2 be two axes that intersect at an angle β and that they rotate at an angular velocity φ_1 and φ_2 (top right image). How can you – except with combined Cardan joints – fulfil the condition $\varphi_1 = \varphi_2$? Let us consider two equally wide cylinders of rotation around the axes. The two corresponding generatrices e_1 und e_2 intersect in the point S. For symmetry reasons, this point lies in the symmetry plane σ of a_1 and a_2, the synchronous plane, and it thus moves on an ellipse.

Torus-shaped grooves and a spherical cage

How can we achieve that S always remains in σ? For this, we need mobile spheres that always remain in σ because they are kept at a distance as they roll along torus-shaped grooves T_1 and T_2, inside a larger sphere Σ_1 and on a smaller sphere Σ_2. They are held together by a (yellow) spherical cage Σ_3, where the spheres have some space to move, as the sphere centres Σ_4 do not move uniformly on a circle in σ.

The great advantage of this constant velocity ball joint is (as with the combined Cardan shaft) that the deflection angle β can be modified freely at any point in time.

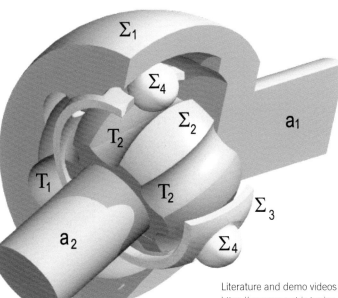

Literature and demo videos
https://www.geometrie.tuwien.ac.at/stachel/087_Gleichlauf.pdf
http://tethys.uni-ak.ac.at/cross-science/spherical-joint.mp4
http://tethys.uni-ak.ac.at/cross-science/spherical-joint-part.mp4

Gearwheels as an alternative

When the angle between the two intersecting axes of rotation is constant in an application, then gearwheels are a good alternative because they are very robust and they allow simultaneous proportional angular velocities.

In the left photo, you can see how a rotation around an axis into an axis is transmitted into a uniform second rotation with a different number of revolution. In the computer animation below (see video), the transmission is $1:1$.

Variation of the transmission ratio via the tooth number

On p. 38 we will discuss how the aperture angles of cones must be changed to achieve (almost) any transmission ratio (the limitation is due to the number of teeth on the wheels).

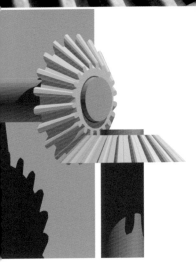

Demo video
http://tethys.uni-ak.ac.at/cross-science/spherical-gears.mp4

Classic Gearwheel

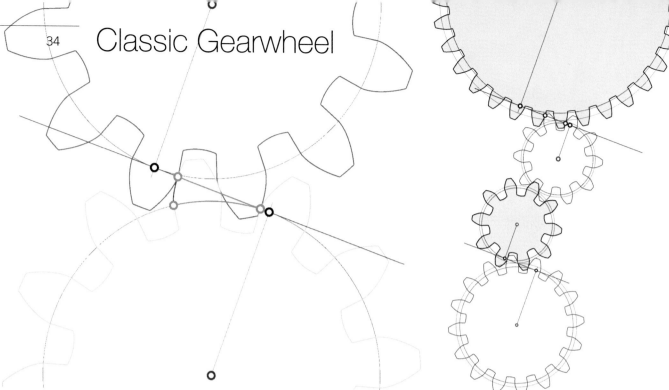

Involute gears

Gearwheel technology has evolved over a century and a half, and there is little point in "reinventing the wheel". While there are several types of gearwheels, involute gears are the most widely known. They have a great variety of applications, and they transmit rotations reliably and exactly from one axis onto a parallel axis. The translation ratio can generally be chosen freely.

Line of action

Without much stress we can imagine two circles (without tooth flanks) that roll on each other without gliding. The respective contact point is known as pitch point. If we add tooth flanks to the circles, then the line of action is the locus of all contact points that occur during the meshing of the gear teeth. For each point of contact, the normal must run on the tooth trace through the pitch point. Theoretically, one tooth trace could be given, and this would then automatically yield the profile of the other trace. With involute gears, both trace types are equivalent, namely so-called involutes of a circle. We can thus produce whole series of gearwheels with a given tooth height and number which mesh perfectly into each other.

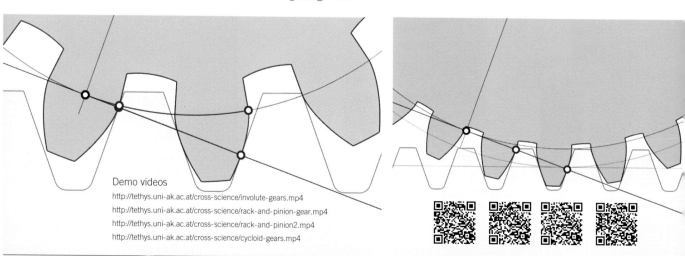

Demo videos

http://tethys.uni-ak.ac.at/cross-science/involute-gears.mp4
http://tethys.uni-ak.ac.at/cross-science/rack-and-pinion-gear.mp4
http://tethys.uni-ak.ac.at/cross-science/rack-and-pinion2.mp4
http://tethys.uni-ak.ac.at/cross-science/cycloid-gears.mp4

Of Clockworks and Planetary Gears

A seeming jumble of gears…

…can be found in every analogue wristwatch or pendulum clock (image on the left and on the right page). Again and again, it is a matter of transforming a constant drive in different ways to produce the different angular velocities of the clock hands.

Planetary gears…

…have many applications in technology, such as in gearboxes, cable winches, and bicycle hub gears. A variation on the principle is also notable (images on the right): If the axes of the planetary gears are fixed, and the outer gear is allowed to rotate, then it rotates significantly slower than the drive shaft.

Demo videos
http://tethys.uni-ak.ac.at/cross-science/clockwork.mp4
http://tethys.uni-ak.ac.at/cross-science/clockwork2.mp4
http://tethys.uni-ak.ac.at/cross-science/planetary-gears1.mp4
http://tethys.uni-ak.ac.at/cross-science/planetary-gears2.mp4

Spherical Trochoids

Relative movement

With conical gearwheels, both cones rotate – in relation to the observer in a fixed system – around their own axis. In relation to one of the two cones (here: to the red one), the other cone (here: the green one) rotates both around its axis and – proportional to this – around the axis of the fixed cone. During this process, points that are connected to the moving cone will move on spheres around the intersection point of both axes. The point

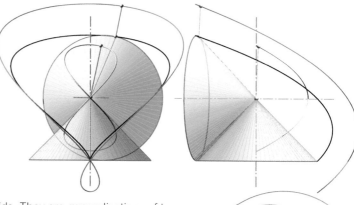

trajectories are thus known as spherical trochoids. They are generalisations of trochoids in the plane, and they commonly appear as such in special projections (the three images on top of the left page show such special views).

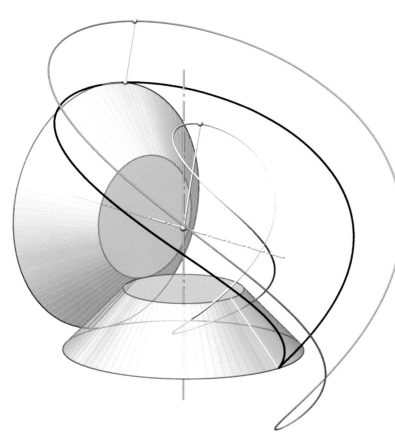

Two parameters

If both cones are contained by a sphere around the intersection point, we can put the radii of its base circles in relation to each other. If the moving green circle must be rolled k times on the fixed red circle in order to return to its starting position, then we can say: the ratio between the relative angular velocities around the cone axes is $1 : k$. The second parameter is the angle α that is enclosed by the two axes. Special values for k and α are obviously particularly interesting. For the example on this page, the simplest case $k = 1$ and $\alpha = 90°$ is illustrated here (see horizontal projection, vertical projection, and ortographic projection views above). The grey point trajectory is especially interesting: it is known as Viviani's curve, which appears as a circle in horizontal projection, as a parabola in vertical projection, and as a figure-eight loop (lemniscate) in orthographic projection.

Demo videos

http://tethys.uni-ak.ac.at/cross-science/cones-rolling1.mp4
http://tethys.uni-ak.ac.at/cross-science/viviani.mp4

$k = 1$
$\alpha = 60°$

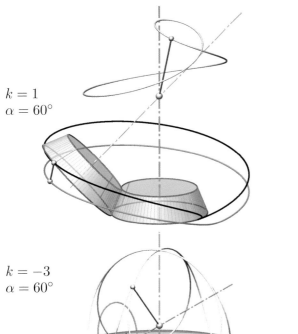

$k = 4$
$\alpha = 60°$

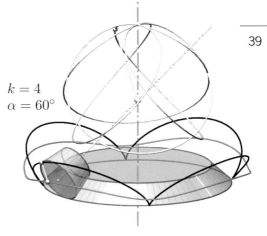

$k = -3$
$\alpha = 60°$

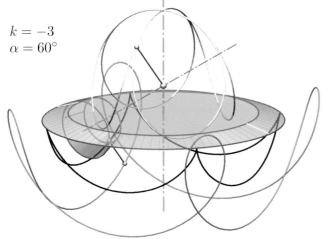

$k = -3$
$\alpha = 60°$ (horizontal projection)

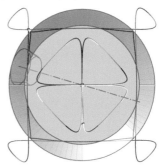

$k = -0.5$
$\alpha = 60°$

Plane rolling on sphere

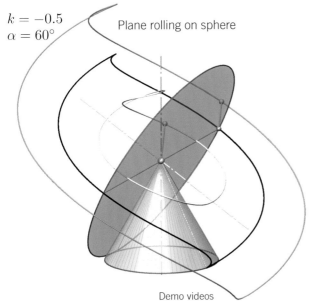

$k = -2$
$\alpha = 60°$

Cone rolling on plane

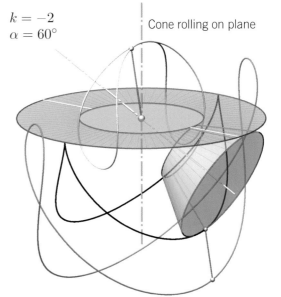

Demo videos

http://tethys.uni-ak.ac.at/cross-science/cones-rolling2.mp4
http://tethys.uni-ak.ac.at/cross-science/cones-rolling3.mp4
http://tethys.uni-ak.ac.at/cross-science/cones-rolling4.mp4
http://tethys.uni-ak.ac.at/cross-science/cones-rolling5.mp4

Patterns and Fractals: Simulation of Nature

Parquets on the Basis of Hexagons

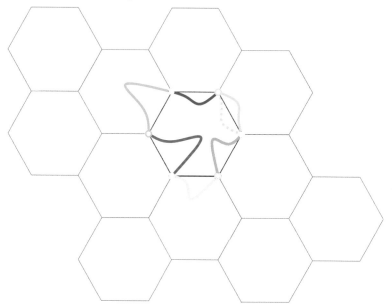

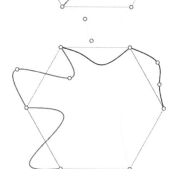

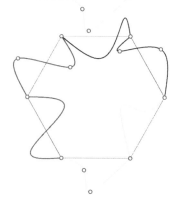

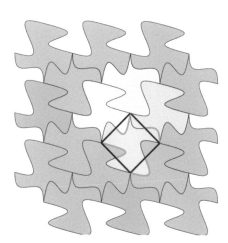

Parquetting with regular hexagons

A plane can be "tiled" seamlessly with regular hexagons. Each tile in this plane can be rotated by any given multiple of 60°.

Here is a trick:

Let us reshape every second side of the hexagon into an arbitrary curve (e.g. blue, green, and yellow in the images). Now we will mirror the three curves on the centre of the hexagon (in the image: bright blue, dashed curve) and rotate them by 60° around the centre (the colours remain the same in the images). We are thus adding or removing what we have previously removed or added. The new component can now be reproduced, and the pieces can be assembled like a jigsaw.

The idea is as ingenious as it is simple…

…and can also be done with squares and rhombi (bottom right images). Many ideas by M. C. Escher are based on this method.

Demo video
http://tethys.uni-ak.ac.at/cross-science/escher-tiles.mp4

Parquets with Semi-regular Pentagons

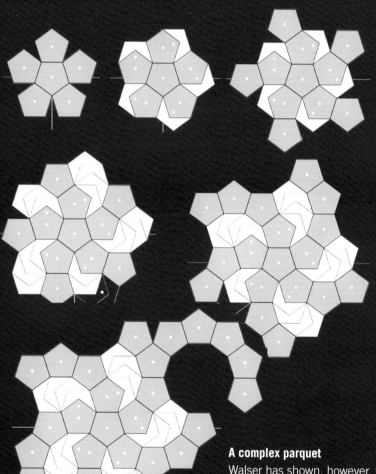

Regular and semi-regular pentagons

We can create a semi-regular pentagon by cutting a rhombus out of a regular pentagon, as in the image to the left. Hans Walser had used this new element to build various ornaments and parquets (a relatively simple example of such an ornament can be seen on the left; three less trivial parquets that are entirely made of semi-regular pentagons are found below).

A complex parquet

Walser has shown, however, that such parquets can be much more complicated. In the image series on the left, you can see to some extent how this pattern is built concentrally, step by step, from a basic pattern made of regular pentagons. The page-filling pattern on the right-hand page is a rectangular section where 1,500 elements have been strung together.

Demo video and theory

http://tethys.uni-ak.ac.at/cross-science/walser-tiling.mp4

Hans Walser **Semi-Regular Figures Between Beauty and Regularity.** *In: C. Michelsen, A. Beckmann, V. Freiman, and U.T. Jankvist (ed.), Mathematics as a Bridge Between the Disciplines, 29–38 (2018)*

https://pure.au.dk/ws/files/137592156/MACAS_2017_Proceedings_Lindenskov_seven_keys.pdf

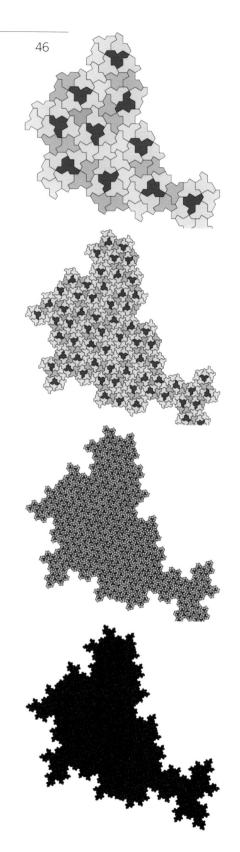

The Einstein Tile

A single prototype to create an aperiodic pattern!

The previous two double pages dealt with periodic and aperiodic patterns. Walser tessellations consist of only two tile types: regular and semi-regular pentagons. The year 2022 saw a breakthrough with a pattern that requires only one single prototile. The news received significant press coverage.

The label "einstein tile"…

…could be taken literally: one stone (one prototile) is enough! The proof for this requires considerable mathematical know-how. It was possible to show that the patterns can be grouped in four cluster types (bottom right image) that can be arranged aperiodically. By combining the clusters, you get structures with the same shape that keep increasing in size. The image series to the left shows growing hierarchies that – in order to fit on this page – keep shrinking.

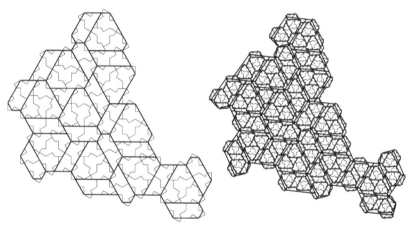

The recursive programming of the pattern soon yields a great amount of building blocks. If we keep shrinking the image, we get a fractal-like structure (p. 49). The authors Myers, Kaplan, and Goodman-Strauss have shown that the 13-sided einstein tile is formed by combining equilateral triangles. They thus discovered an entire family of aperiodic "polydiamond tiling".

Theory and demo videos

https://www.spektrum.de/news/hobby-mathematiker-findet-lang-ersehnte-einstein-kachel/2124963

https://static.spektrum.de/fm/976/animation.7828920.gif

D. Smith, J. S. Myers, C.S. Kaplan, C. Goodman-Strauss

An aperiodic monotile. *Preprint (2023)*

https://arxiv.org/pdf/2303.10798.pdf

http://tethys.uni-ak.ac.at/cross-science/einstein-tiles.mp4

Mandelbrot and Julia Sets

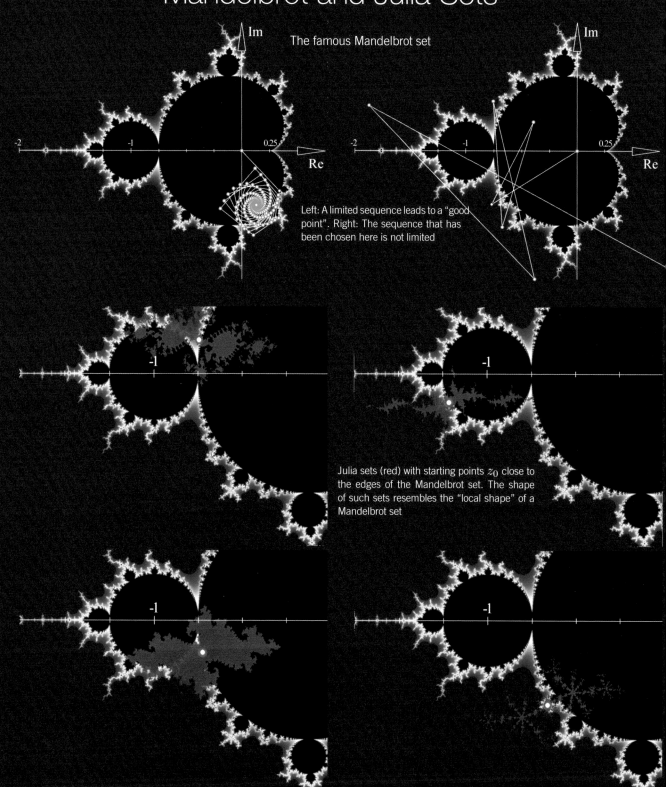

The famous Mandelbrot set

Left: A limited sequence leads to a "good point". Right: The sequence that has been chosen here is not limited

Julia sets (red) with starting points z_0 close to the edges of the Mandelbrot set. The shape of such sets resembles the "local shape" of a Mandelbrot set

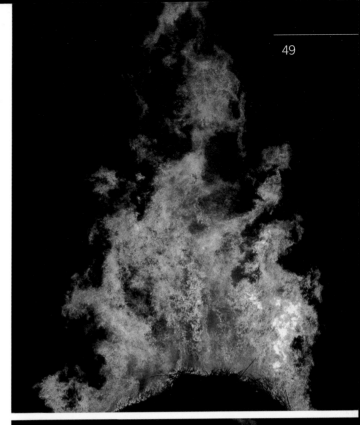

The Mandelbrot set

It was only in the 1980s that fractals achieved their modern-day popularity with the astonishing images by Benoît Mandelbrot. The appearance of the first graphics-capable computer systems enabled the aesthetically fascinating visualisation of hitherto unknown fractal worlds. We look at the complex function $f(z) = z^2 + c$ and then define a sequence of complex numbers $z_0, z_1, z_2, \ldots$ as follows:

$$z_0 = 0$$
$$z_1 = f(z_0) = c$$
$$\vdots$$
$$z_{n+1} = f(z_n) = z_n^2 + c.$$

When the sequence z_n remains finite (top left image on the left-hand page), then we consider the value c to be "good" and we mark it with a black or blue point in the Gauss plane. Mandelbrot sets can thus be defined as the set of all values c for which the recursion $z_{n+1} = z_n^2 + c$ remains finite when you choose $z_0 = 0$.

Julia sets

Julia sets (named after Gaston Julia, who studied these sets several decades before the computer age) can be generated with the same recursive formula as Mandelbrot sets. Instead of $z_0 = 0$, z_0 can be any point on the Gaussian plane. However, while the Mandelbrot set is the set of complex numbers c for which the iteration remains finite, the Julia set is the set of the iterations z_n – or more precisely, the edge of it. If you choose, for instance, a point z_0 along the margins of the Mandelbrot set, then the shape of the corresponding set will change quickly.

The images to the right illustrate snapshots of a solstice bonfire, where the fractal nature of flames is clearly visible. The photographs resemble the – purely theoretically generated – Julia sets on the left-hand page.

Demo video

http://tethys.uni-ak.ac.at/cross-science/mandelbrot.mp4

A Glimpse into Barnsley's Herbarium

First we need four linear transformations

Take a point with the coordinates (x, y) and apply a linear transformation onto it:

$$f(x, y) = \begin{bmatrix} a & b \\ c & d \end{bmatrix} \begin{bmatrix} x \\ y \end{bmatrix} + \begin{bmatrix} e \\ f \end{bmatrix}$$

The matrix operation will yield the coordinates of a new point. Now let us choose "suitably" four such operations and repeatedly apply one of these four operations at random. We thus keep getting new points of the plane.

What may look like just fun and games will soon solidify (in the truest sense of the word) into a structure branching into multiple directions that – depending on the given coefficients – is often reminiscent of plant structures like ferns, for instance.

Self-similar mathematically generated patterns

Remarkably, the result is a fractal that is strongly reminiscent of plant structures. In nature you can, in fact, find such variations from one reduction stage to the next; no two reduction stages are exactly the same. However, in nature the fractal principle often ends already with the second reduction stage (top right photograph). With ferns, for instance, you will then find the spores of the plants. Overall, the plant thus achieves an optimal distribution of spores.

The video cited below shows how, with a computer, we can rapidly generate new – mostly organic-looking – structures by varying the coefficient of the matrix. You can also "play around" with the accompanying software.

Demo video
http://tethys.uni-ak.ac.at/cross-science/barnsley-fern.mp4

How Do Ferns Grow?

The gaps between the branching points form more or less an arithmetic sequence: each gap is created from the previous one by summing a constant. So, after a certain number of points, the distance becomes zero.

When developing or transforming, we rotate step by step at each branching point by the same angle. The resulting winding curve is approximately a clothoid, also known as Euler spiral.

Demo videos
http://tethys.uni-ak.ac.at/cross-science/ferns-growing.mp4

As close as possible to nature

The computer simulation takes into account as much as possible of what can be observed during the growth of ferns. The size generally increases in a linear manner. The fractal behaviour is – as in nature – interrupted after the second stage. Each branching mini fern also grows linearly and uncurls simultaneously and analogous to the main stem.

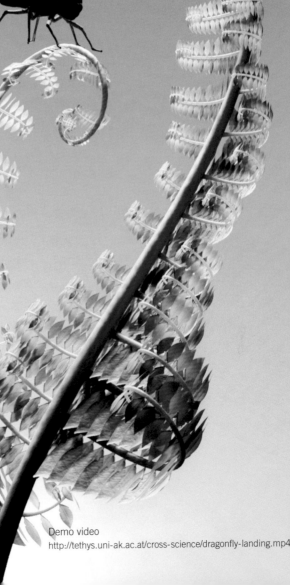

Demo video
http://tethys.uni-ak.ac.at/cross-science/dragonfly-landing.mp4

Fractal Formations Made of Small Spheres

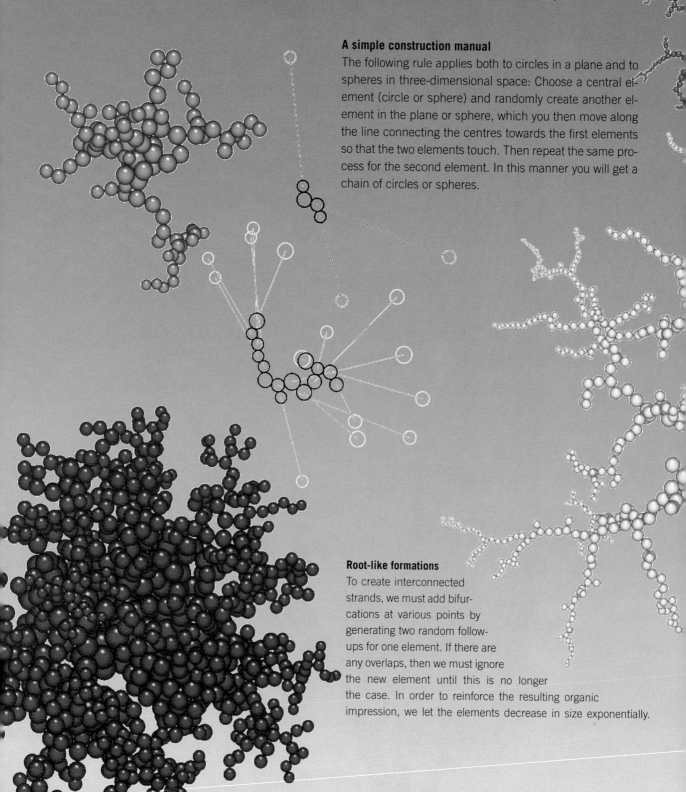

A simple construction manual

The following rule applies both to circles in a plane and to spheres in three-dimensional space: Choose a central element (circle or sphere) and randomly create another element in the plane or sphere, which you then move along the line connecting the centres towards the first elements so that the two elements touch. Then repeat the same process for the second element. In this manner you will get a chain of circles or spheres.

Root-like formations

To create interconnected strands, we must add bifurcations at various points by generating two random follow-ups for one element. If there are any overlaps, then we must ignore the new element until this is no longer the case. In order to reinforce the resulting organic impression, we let the elements decrease in size exponentially.

Hard coral with seahorse

To the right you can see a hard coral in which – revealed by a closer look – a perfectly camouflaged pygmy seahorse lives. In this case, the approximation through a fractal does not require a shrinking of the spheres.

Demo videos
http://tethys.uni-ak.ac.at/cross-science/nearest-circle2d.mp4
http://tethys.uni-ak.ac.at/cross-science/nearest-sphere.mp4

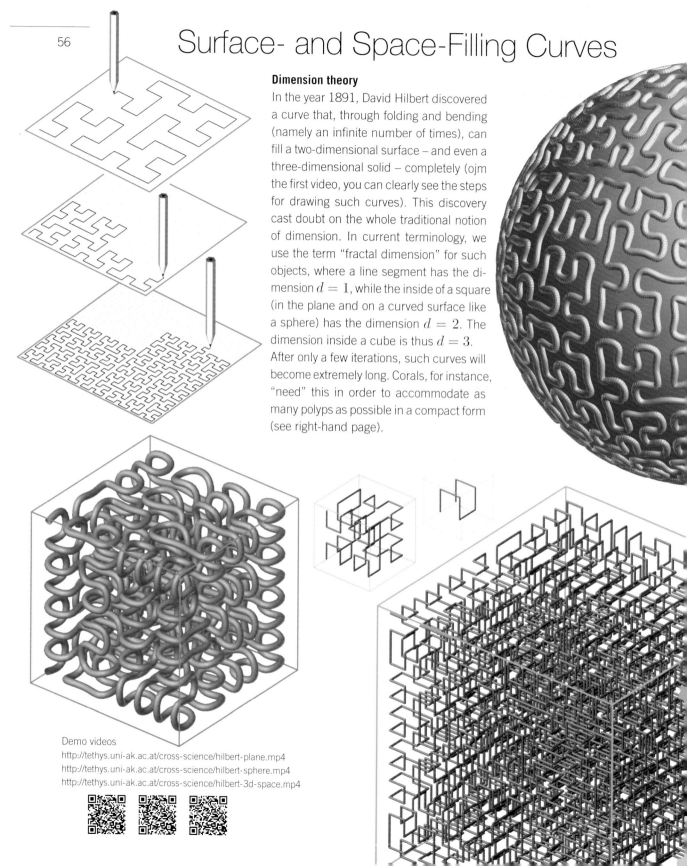

Surface- and Space-Filling Curves

Dimension theory

In the year 1891, David Hilbert discovered a curve that, through folding and bending (namely an infinite number of times), can fill a two-dimensional surface – and even a three-dimensional solid – completely (ojm the first video, you can clearly see the steps for drawing such curves). This discovery cast doubt on the whole traditional notion of dimension. In current terminology, we use the term "fractal dimension" for such objects, where a line segment has the dimension $d = 1$, while the inside of a square (in the plane and on a curved surface like a sphere) has the dimension $d = 2$. The dimension inside a cube is thus $d = 3$.

After only a few iterations, such curves will become extremely long. Corals, for instance, "need" this in order to accommodate as many polyps as possible in a compact form (see right-hand page).

Demo videos

http://tethys.uni-ak.ac.at/cross-science/hilbert-plane.mp4
http://tethys.uni-ak.ac.at/cross-science/hilbert-sphere.mp4
http://tethys.uni-ak.ac.at/cross-science/hilbert-3d-space.mp4

Mathematically Generated Fur Patterns

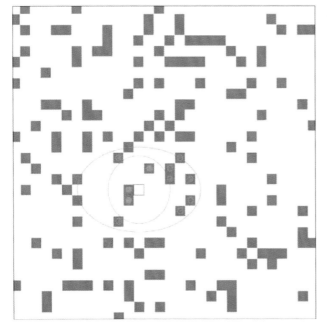

A pixel grid

We choose a grid of, say, 100×100 points and randomly colour a number of pixels black ("pixel" is short for picture element, referring to a small square in a grid). Now we move systematically over the grid, pixel by pixel (in the image to the left, one such test pixel has been marked red).

Concentric rings

Now we imagine two (elliptical or circular) rings around the test pixel, with the outer (orange) ring being about twice as large as the inner (green) one.

A counting procedure and a subtraction

A simple counting process starts now: we count the black pixels that lie in the plane that is bounded by the inner and outer ring (marked orange, amount n) as well as the black pixel that lie inside the inner ring (marked green, amount m). If, for instance, $n > 3m$ (or $n-3m > 0$), then the test pixel will temporarily turn black. After testing all the pixels, the patterns will have changed.

Quick convergence of the pattern

Repeating this process will yield a new image, but lo and behold: The pattern will quickly approximate its final look, which becomes recognisable after only 5 or 10 iterations. The weighting (multiplication) by the factor 3 is because there can be no more than about three times as many orange pixels ("inhibitors") as green pixels ("activators"). If there is a predominance of weighted activators, then the test pixel will turn black.

The random choice of a starting point is barely relevant

The four computer-generated "zebra patterns" that can be seen on the left-hand page were generated in the manner outlined above. Surprisingly, it is not the number or position of starting points that determine the shape of the pattern, but rather the form of the two rings (in order to get zebra patterns, you must choose two ellipses like the ones in the top left sketch: their main axes are separated by $90°$).

The photo at the bottom of the left-hand page shows a zebra mother with its baby. If you compare the patterns on their heads, you will spot strong similarities due to their close relation. Comparable patterns can be found not only in animal furs and skins (tigers, tiger sharks) but also in the sand ridges of shallow waters (image to the right).

Elliptical or circular rings?

If the area is formed by two concentric circles, there is no longer an ideal direction for the resulting pattern. Then, the images converge towards irregularly distributed spots reminiscent of the fur patterns of cheetahs or leopards (photo below left). A corresponding selection of parameters is shown in the images on the bottom right. The size of the concentric rings and their weighting w when calculating the difference are quite delicate and determine the number of dark spots.

$14 - w*46 = -4.4000$

$9 - w*33 = -4.2000$

Demo videos
http://tethys.uni-ak.ac.at/cross-science/zebra.mp4
http://tethys.uni-ak.ac.at/cross-science/leopard.mp4

Strange Angles: Spirals?

Of Sunflowers and Daisies

Spirals?

When you look long enough at a sunflower (top image) or a common daisy (left image) from above, you will inevitably discover some spirals in their patterns. You might even recognise two counter-rotating spirals. By counting the number of spirals in a bundle, we will realise that they are always so-called Fibonacci numbers (more on that on the next page). This famous number sequence is closely linked to the *golden ratio*, as well as to exponential growth. Are sunflowers and common daisies familiar with the golden ratio, and do they grow – as opposed to most plants – exponentially?

Points on virtual spirals

On the next double page, we will discuss a principle that shall be outlined here in a simplified manner: Let us choose a centre and a nearby starting point. Then we copy this point and rotate the copy by the golden ratio around the centre, increasing the distance from the centre exponentially. This process can be repeated any number of times (first video). The individual points thus lie on so-called logarithmic spirals.

The Fibonacci sequence

During this process the distances will increase almost exactly by a ratio of $1 : 2 : 3 : 5 : 8$ etc. This series of numbers is actually quite famous: it is known as the Fibonacci sequence. The next number in the series is always found by adding the two previous numbers together (e.g. $3 + 2 = 5$ and $5 + 3 = 8$).

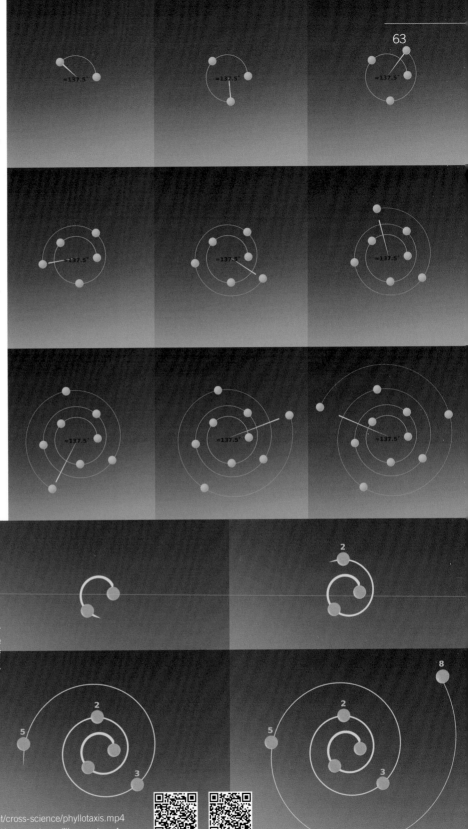

In the second video, you will see how – starting from the first point – you can move in the opposing direction, but instead of the rotational angle, we use a 360° complementary angle.

Demo videos
http://tethys.uni-ak.ac.at/cross-science/phyllotaxis.mp4
http://tethys.uni-ak.ac.at/cross-science/fibonacci.mp4

A Genetically Predetermined Angle?

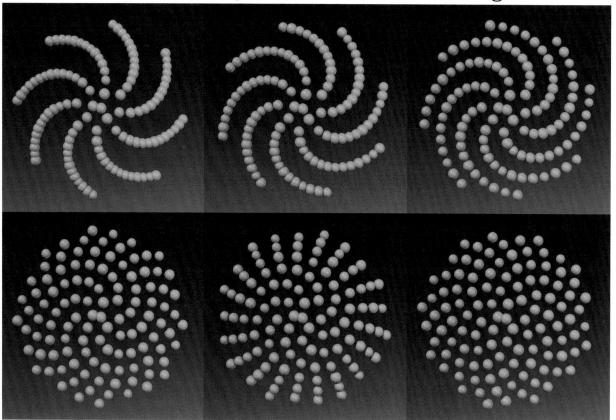

An optimal angle of rotation

From the common daisy to the metre-high sunflower – they all follow the *golden ratio* γ, which is linked to the *golden number* $\Phi = (1 + \sqrt{5})/2$:

$$\gamma = (\Phi - 1) \cdot 360° \approx 137,5°.$$

These flowers "need" the golden ratio in order to accommodate as many seeds as possible on a circular surface.

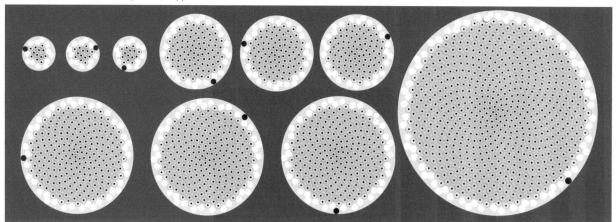

Demo videos

http://tethys.uni-ak.ac.at/cross-science/spiral-worse.mp4
http://tethys.uni-ak.ac.at/cross-science/spiral-best.mp4
http://tethys.uni-ak.ac.at/cross-science/sunflower.mp4
http://tethys.uni-ak.ac.at/cross-science/phyllotaxis2.mp4

A simple computer model

The following model has been used on the left-hand page: Start with the first seed right next to the centre of the circle and place the subsequent seeds by rotating by the golden ratio and moving "a tiny bit" away from the centre. By calibrating the distance parameter a little, you will get images that are astonishingly reminiscent of sunflowers. One might think that it does not matter whether we always rotate by $137,5°$ or e. g. by one or two degrees less. However, this is by no means the case as the image series on the left-hand page shows: in these images, the rotation deviates from the golden ratio by only one or two degrees at the maximum.

Plants have found the golden ratio through evolution

A sunflower propagates through its seeds. These are usually picked by birds and dispersed because the birds will occasionally drop a few seeds. One thing is clear: the more seeds, the greater the probability of long-term survival as a species.

Now imagine a plant species (a predecessor of our modern sunflower) that has adopted the following growth strategy in the course of evolution (because the strategy has proven to be advantageous): "place your seeds in such a manner that you first produce a seed close to the centre and then you arrange each successive seed by rotating the direction far enough and also moving slightly away from the centre for an ideal use of the available space."

As opposed to sunflowers, common daisies distribute their seeds on curved surfaces (spherical or conical). Both plants tend to develop their seed heads under cover because their petals are closed in the process.

Demo video
http://tethys.uni-ak.ac.at/cross-science/plant-opening.mp4

Optical Illusion

The visible spirals are an optical illusion

The algorithm for the creation of new single flowers does not produce one seed after another along a spiral, but instead it keeps adding new seeds at the margin, namely in line with the golden ratio. With optimal packing you will get optical illusions featuring two series of spirals – one in clockwise direction and the other in counterclockwise direction. Close to the centre, these spirals have several kink points.

The intensity of the illusion can vary

In the large image below you can see that the illusion is strongly dependent on the form of the flower bud: If these are packaged in a hexagonal manner – like a honeycomb, but growing from the inside to the outside – then the illusion will disappear. Flower buds that are shaped like a deltoid, on the other hand, reinforce the illusion, and the aforementioned spirals are dominant.

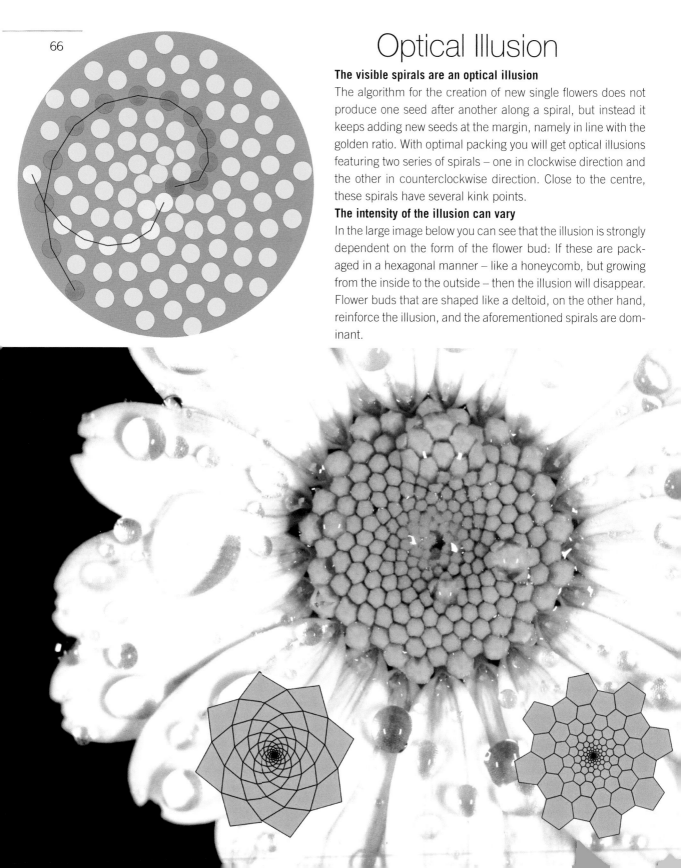

No spirals up close

In close-ups like the image to the right, there are no visible spirals. You can see: The crucial thing here is dense packing to accommodate as many seeds as possible. The burst solitary flower on the right measures only a millimetre, but it is still reminiscent of a lily's calyx, which is a hundred times larger.

Dense packing

In the bottom right photograph, the base on which the seeds are arranged has almost the form of an ellipsoid of revolution, also known as spheroid. The seeds on its surface are arranged according to a principle that is similar to that on a plane, which can be shown through computer simulation. Nature will ultimately stretch and squeeze to make optimal use of the available volume. If you magnify the tiny spheres used in the simulation, you will get images like the ones at the bottom left, which actually look quite realistic. In these images you will keep stumbling across more alleged spirals.

Demo videos
http://tethys.uni-ak.ac.at/cross-science/phyllo-ellipsoid.mp4
http://tethys.uni-ak.ac.at/cross-science/phyllotaxis3.mp4

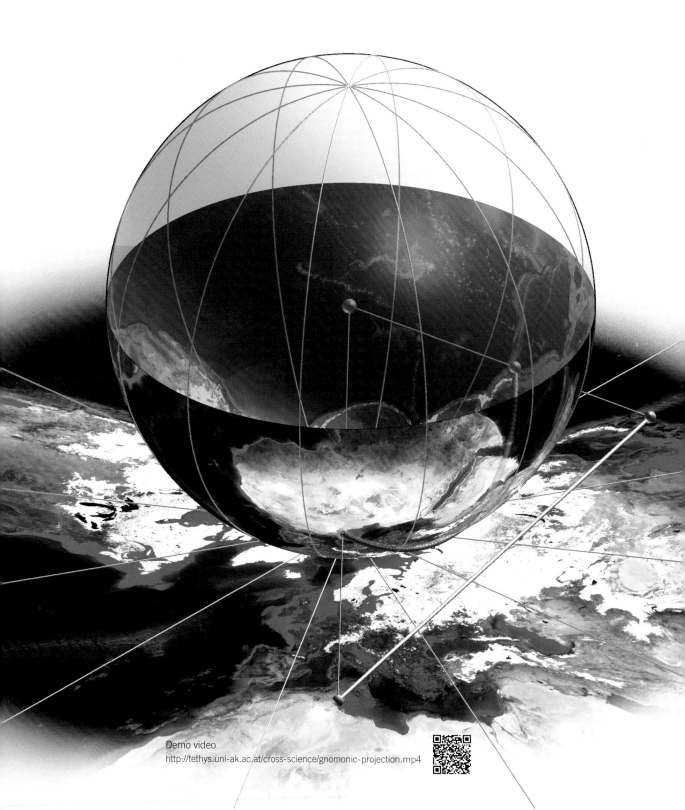

Demo video
http://tethys.uni-ak.ac.at/cross-science/gnomonic-projection.mp4

Projections:
Necessary and Practical

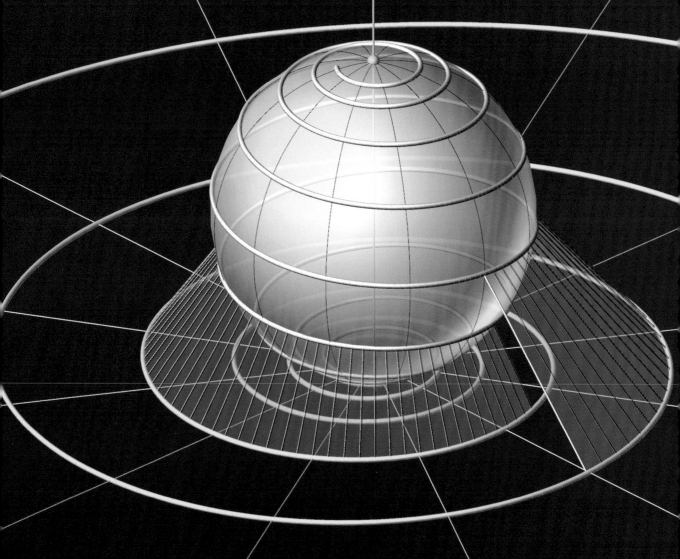

Demo video
http://tethys.uni-ak.ac.at/cross-science/spiral-stereographic.mp4

Equirectangular Projection – Pros and Cons

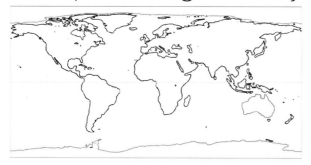

Standard representation of the Earth's surface

The surface of the Earth is very often represented as a rectangle in a Cartesian (λ, φ)-coordinate system (with λ being the geographical longitude and φ the geographical latitude). In these maps the circles of longitude and the circles of latitude appear as straight lines. For historical reasons the prime meridian runs through Greenwich in London. At first this might seem like a sensible way of representing the Earth because you can draw and find specific spots on the Earth's surface very easily.

Tearing continents apart

Along the map margins $\lambda = -90°$ and $\lambda = +90°$, the continents are "torn apart", which is not noticeable on the standard world map because there are no continents in this area (top left on the right-hand page). However, if we choose a different prime meridian that runs, for instance, through New Zealand instead of Greenwich, then Europe and Africa are torn apart (bottom left on the right-hand page).

Extreme distortions

Almost worse than the aforementioned: The map becomes increasingly distorted as we move further away from the equator. Greenland with its 2 million km^2 thus appears far too big (and wide) compared to Africa (the surface of Africa is a whopping 14 times larger). The extreme case can be found at the Poles, whose corresponding circles of longitude have a radius of zero, but on the map they appear to extend from the very left to the very right edge, so that one might be misled to believe that these circles of longitude are just as long as the equator (40 000 km). The ice surfaces of the North and South Pole thus appear bizarrely distorted.

A comparison of Antarctica and Australia

In maps of the world with classical equirectangular projection, a comparison of the two continents (coloured orange and green in the images) is clearly not possible. One could set, for instance, the difference in the map's circles of longitude in relation to Australia, but Antarctica would then be torn apart (top right on the right-hand page). It is best to place the coordinate origin somewhere between the two continents (bottom right on the right-hand page) and thus minimise the distortion. And, lo and behold, one can see what an incredible match the two continents are – about 80 million years ago, they were, in fact, one unit. With "classical" equirectangular projection, a similar effect can be observed quite well between Africa and South America, because the two continents are relatively well centred.

Satellite orbits over the Earth's surface

The two images below (NASA) show satellite orbits in space (circles around the Earth's centre) and their projection onto an equirectangular world map – on such a map, their curves appear as "phase-shifted waves" (not sinus curves!). The phase shift Δ is due to the Earth's rotation around its own axis by about $22°$ after an orbital period of almost ninety minutes.

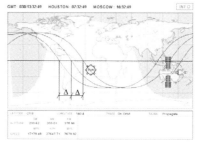

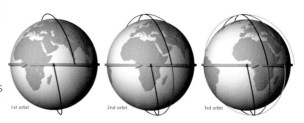

Satellite video (NASA)
https://earthobservatory.nasa.gov/ContentFeature/OrbitsCatalog/images/sun-synchronous_orbit.h264.mov

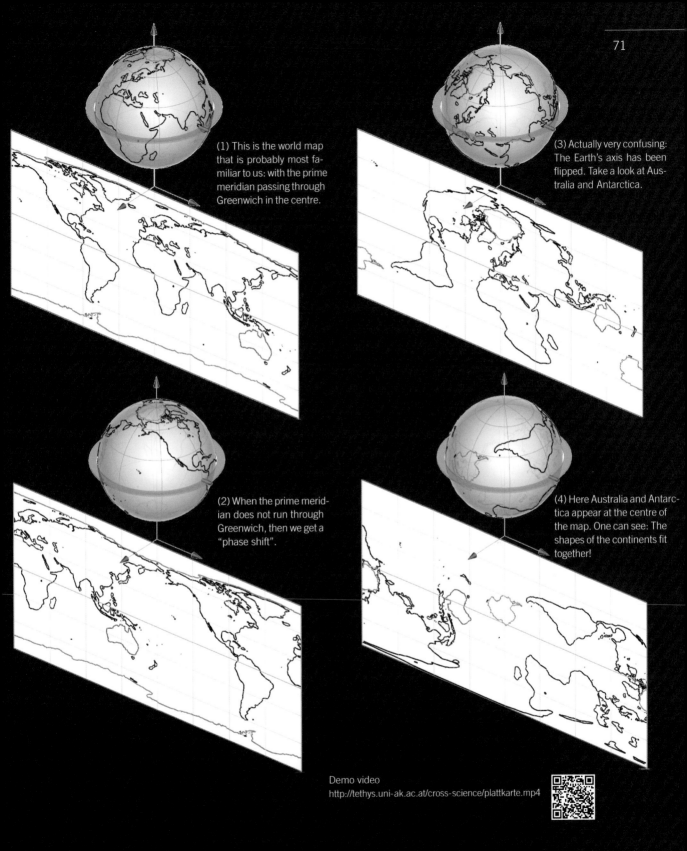

(1) This is the world map that is probably most familiar to us: with the prime meridian passing through Greenwich in the centre.

(3) Actually very confusing: The Earth's axis has been flipped. Take a look at Australia and Antarctica.

(2) When the prime meridian does not run through Greenwich, then we get a "phase shift".

(4) Here Australia and Antarctica appear at the centre of the map. One can see: The shapes of the continents fit together!

Demo video
http://tethys.uni-ak.ac.at/cross-science/plattkarte.mp4

Thermohaline Circulation

Twenty million cubic metres of saltwater per second

This enormous amount (slightly more than half of the existing freshwater) flows per second at often great depths with a velocity of 1 to 3 km per day, moving in loops multiple times around the Globe. At certain places, e.g. in the Gulf of Mexico, this stream "gets caught up", heats up, and rises, reaching the surfaces in polar regions relatively fast (Gulf Stream!). There the stream cools down significantly and sinks like a colossal waterfall. In thus travels an estimated 600 km per year and takes about 200 years for a full circulation.

The stream circumnavigates Antarctica several times. If we now use an equirectangular projection map, like many publications do, then the ribbon will "tear" as in the bottom right image. This makes it very difficult to visualise things – so, in this case only a projection from the North Pole or a "view from below" would help here (bottom left image).

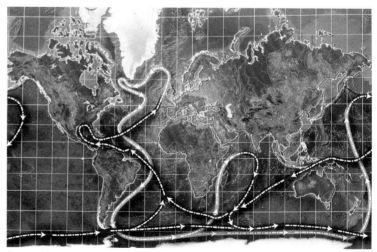

Demo videos
http://tethys.uni-ak.ac.at/cross-science/gulfstream.mp4
http://tethys.uni-ak.ac.at/cross-science/global-conveyor-belt.mp4

As many images as possible…

…are needed so we can say that we have understood the complicated loops with their supposed overlaps. It helps if we colour the deep cold-water portions blue while marking the warm portions close to the surface red. Ideally, we also have access to some 3D animations.

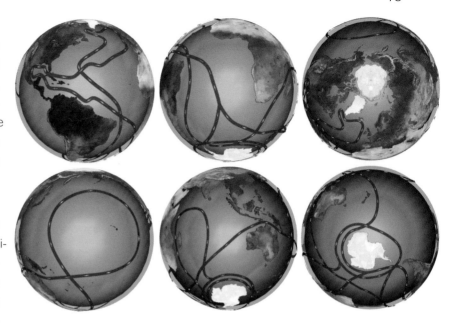

Why do the currents turn?

Due to the Earth's rotation, there is an additional apparent force (the Coriolis force): A current – be it air or water – that pushes from the north to the south is redirected to the west on the northern hemisphere (because it does not carry enough rotational energy to keep up with the Earth's rotation). If the current travels from south to north, it will outpace the Earth's rotation and travel towards the north-east (e. g. like the Gulf Stream). On the southern hemisphere, these relations are reversed. The images on the right show the creation of a hurricane: a low-pressure area sucks in air from the north and south. This generates the vortex around the "eye of the hurricane". On the northern hemisphere, hurricanes spin in a counterclockwise direction, on the southern hemisphere in a clockwise direction.

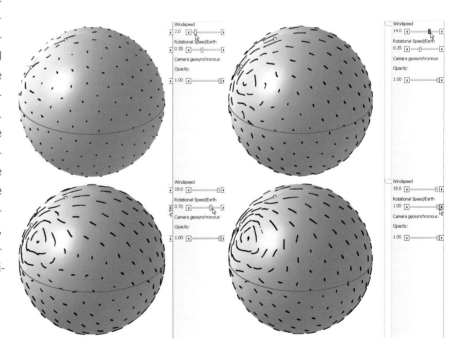

Demo video
http://tethys.uni-ak.ac.at/cross-science/coriolis.mp4

Circle-Preserving Stereographic Projection

Projection from a sphere point

When you project from a point of a sphere onto a plane that is perpendicular to the tangential radius, then you can prove that this projection is "circle-preserving": Circles that are located on the sphere are also circular in the image (exceptions are circles that run through the projection centre: they appear as straight lines).

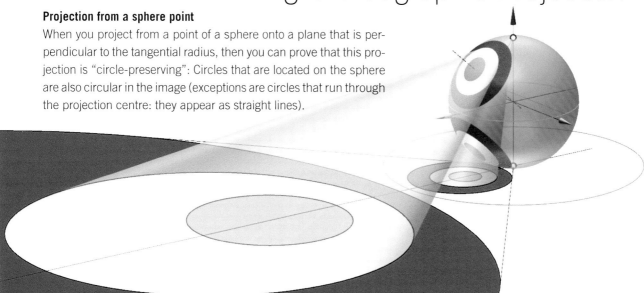

Angle preservation also holds

Another characteristic of this projection is the "preservation of angles": When two curves on the sphere intersect at an angle α, then the corresponding image curves will also intersect at an angle α. A typical example of this can be found in the image on p. 69: If you want to find the curve on the sphere that intersects all meridian circles at the same angle, then you start from the plane. There you will find curves that intersect a bundle of rays through a fixed point at a constant angle, the so-called logarithmic spirals. If you project such a spiral stereographically onto a sphere, then the exercise is solved.

A nice correlation with inversions on a circle

The images below illustrate the correlation between a planar conformal transformation, inversion, and the situation on a sphere. Pre-image and image of the inversion correspond to figures that are symmetrical in relation to the equator of the sphere.

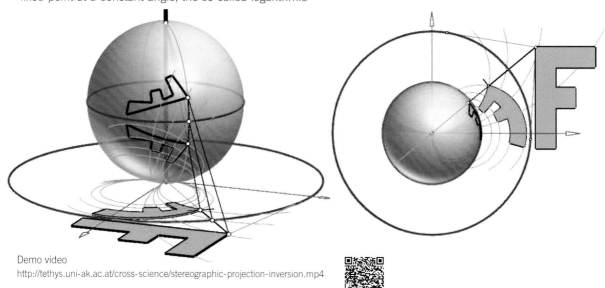

Demo video
http://tethys.uni-ak.ac.at/cross-science/stereographic-projection-inversion.mp4

Projection from the antipodal point

Each point on the Earth's surface has "its" direct opposite. If you project from this antipode of a point to the tangential plane of the first point, you will get a stereographic projection with the advantage that the angles of the intersecting curves of the spheres are preserved in the projected image (which is important, for instance, in navigation and aviation). Moreover, the grid of latitude and longitude circles is then pictured as a circular grid. The simplest case is projecting from the North or South Pole onto the tangential plane in the opposite Pole (top right image). The large image below shows the general case.

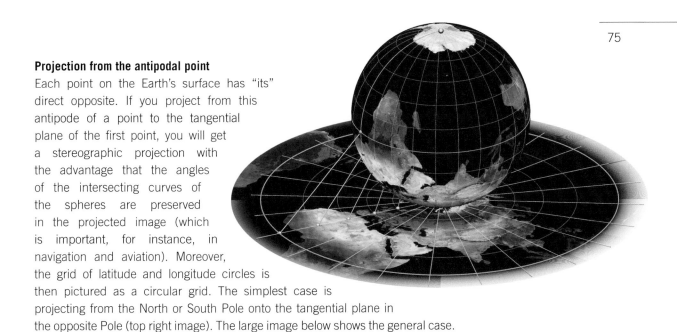

Demo video
http://tethys.uni-ak.ac.at/cross-science/stereographic-antipode.mp4

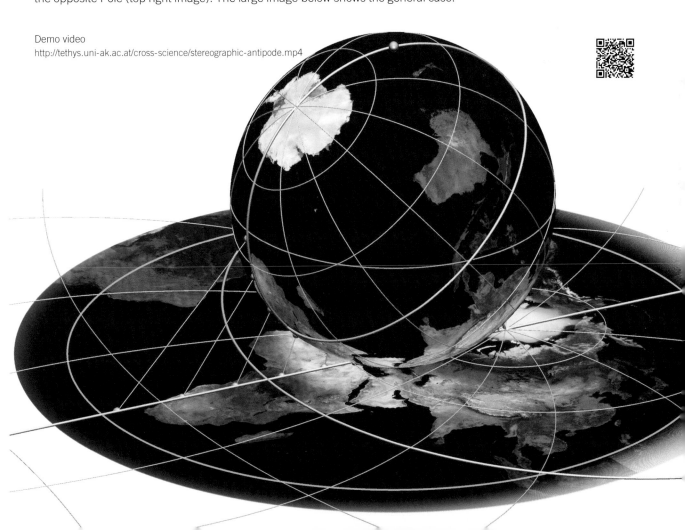

The Sphere Is Rolling…

Normally…

…a (homogenous) sphere can roll in any direction on a horizontal plane. This is so because the potential energy is the same in all positions. If the sphere has grooves that specify an exactly defined rolling motion, as in the image series above, then differential geometry comes into action: now we transfer arc lengths and angles.

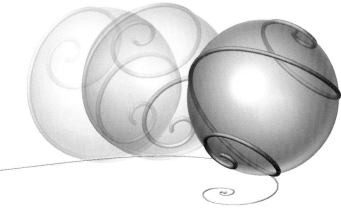

Demo videos
http://tethys.uni-ak.ac.at/cross-science/rolling-spiral2.mp4
http://tethys.uni-ak.ac.at/cross-science/rolling-spiral.mp4

Indentations

In the concrete case on the left-hand page, we take a sphere's rhumb line that intersects the meridian circles at a constant angle. The exactly calculated indentation often resembles logarithmic spirals. The exact calculation requires certain integrals due to the arc lengths.

Scarab beetles...

...want to remove the dung on which the hatched larvae are feeding from danger zones as fast as possible (ideally along a straight line). To achieve this, they roll the dung with stunning precision into spheres that are larger than the beetle itself. Although the rolling of the spheres is re-

peatedly interrupted by obstacles, the beetles can mostly maintain the same direction by orienting themselves toward the Sun.

When they launch into flight...

...it is remarkable as well how scarabs orient themselves toward the Sun (as can be seen from the symmetrical shadow) in a statistically significant manner. For more information on this, see also the video cited below, where a beetle eventually heads in the direction of the Sun. Scarabs' sphere rolling in the direction of the Sun as well as their launching into flight toward the Sun might have earned them their cult status in Ancient Egypt.

Demo videos
http://tethys.uni-ak.ac.at/cross-science/scarab-rolling.mp4
http://tethys.uni-ak.ac.at/cross-science/scarab-rolling2.mp4
http://tethys.uni-ak.ac.at/cross-science/scarab-flying-off.mp4

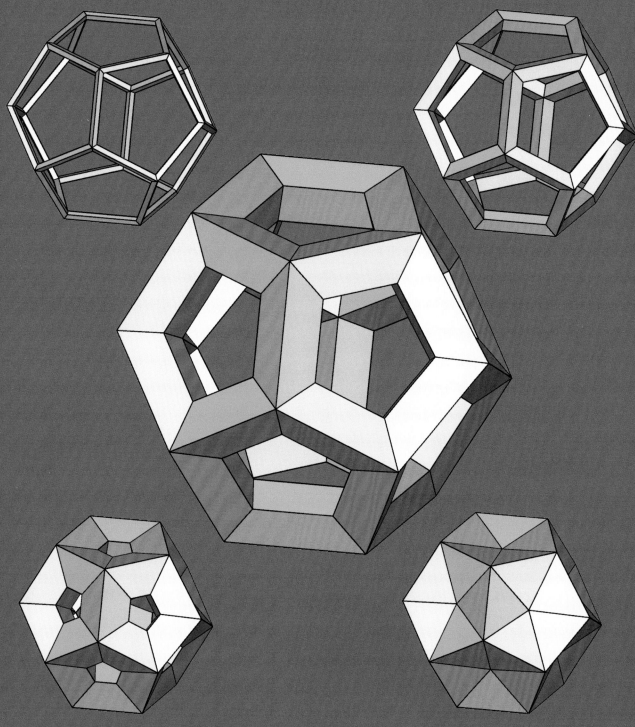

Demo video
http://tethys.uni-ak.ac.at/cross-science/dodecahedron-frame.mp4

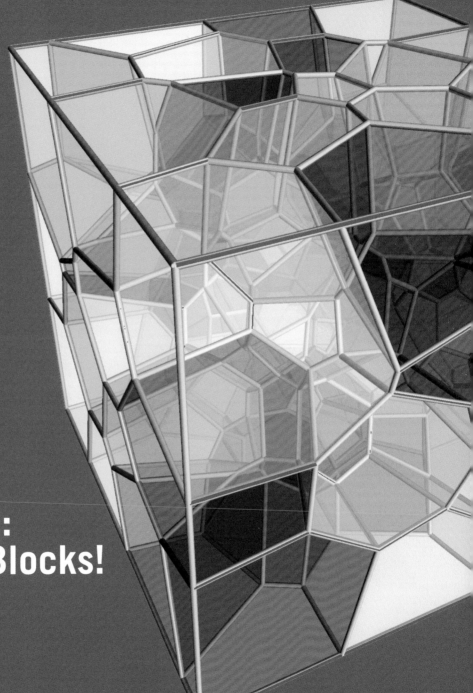

Polyhedra:
Building Blocks!

Demo video
http://tethys.uni-ak.ac.at/cross-science/cubic-puzzle.mp4

New Objects through Dualisation

Special polyhedra and concentric spheres

Polyhedra have vertices, edges, and faces. Special polyhedra are often linked to concentric spheres (inscribed spheres or circumscribed spheres). It thus makes sense to subject them to a special geometric transformation: "polarising" them onto such a sphere. As we will see shortly, the result is a "dual polyhedron" that is easy to identify.

The polarisation process

In this transformation, the vertices of the polyhedron correspond to the supporting planes of the dual polyhedron's faces, and vice versa. The edges of one polyhedron correspond to the edges of the other. The transformation instructions for polarisations onto a sphere (centre M, radius r) state:

The plane π, which corresponds to a point P, is perpen-

dicular to MP and has a distance of $r^2/\overline{MP}$. On the other hand, the point P, which corresponds to a plane π, lies on the normal to π through M at a distance of $r^2/\overline{M\pi}$. Straight lines are transformed by polarising two planes through them and searching for the connection between the corresponding points.

Polarisation of Archimedean solids

The graphic below shows the polarisation of eight Archimedean solids onto a sphere around the centre. Six (blue) Archimedean solids change into (yellow) solids with congruent triangles. Two (grey) Archimedean solids change into solids with congruent rhombi (turquoise). The four brown Archimedean solids yield polyhedra (green) with congruent pentagons or squares.

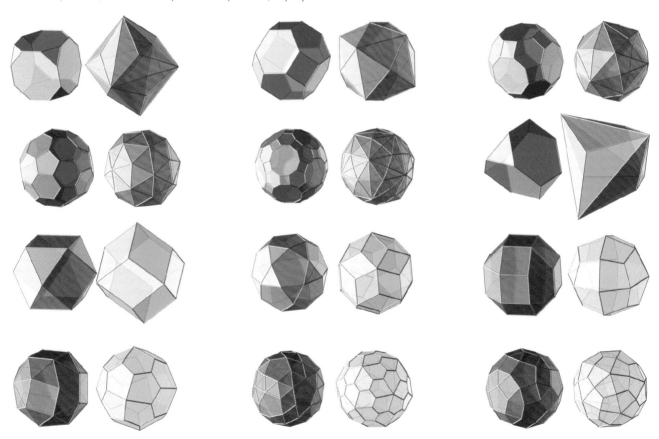

Once again a pentagonal dodecahedron

Let us look at a well-known example: a dodecahedron has twelve regular pentagons and thus $f = 12$ faces, $e = 20$ vertices, and $k = 30$ edges. As a Platonic solid, it has both an inscribed sphere and a circumscribed sphere onto which it can be polarised. We choose the inscribed sphere, which touches the dodecahedron in twelve points, namely the centres of the pentagons.

After polarising the dodecahadron, the twelve faces change into $e^* = f = 12$ vertices of the dual solid. The new solid has $f^* = e = 20$ vertices and $k^* = k = 30$ edges. So, the result is an icosahedron!

Rotating or scaling the pentagons, cutting off the vertices

You can now do some experimenting, and through various procedures on the dodecahedron, you will arrive at Archimedean solids (they carry two or three types of regular polygons). Their dualisation will yield remarkable solids that are entirely covered with congruent facets (usually 60 of them): deltoids, rhombi (here we have only 30), more general pentagons, as well as equilateral triangles. However, they are not Platonic or Archimedean solids, but so-called Catalan solids.

Not quite as perfect but still remarkable

Catalan solids are covered with congruent polygons, but they do not fulfil the strict criteria that apply to Platonic solids. They are still anything but trivial, and it is possible to classify them, including their duality to many Archimedean solids.

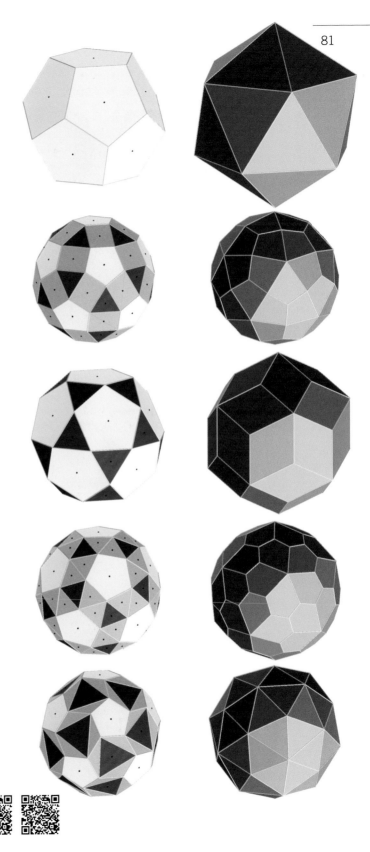

Demo videos

http://tethys.uni-ak.ac.at/cross-science/dualization.mp4

http://tethys.uni-ak.ac.at/cross-science/dodecahedral-dice.mp4

Cutting and Cropping

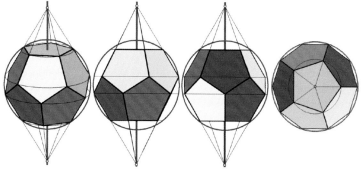

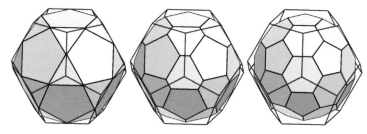

Platonic and Archimedean solids

The five Platonic solids comprise the regular forms known as tetrahedron, hexahedron (cube), octahedron, dodecahedron, and icosahedron. The dodecahedron (also more correctly known as the pentagonal dodecahedron, images at the top) has a few surprises in store. It is dual to the icosahedron (see right-hand page), and it can yield two special solids if you cut the vertices in a certain manner (the left and the right polyhedron in the images below): they are Archimedean solids where two types of regular polygons can be found on the surface.

A first non-trivial spatial tessellation

If you try to stack pentagonal dodecahedra, it will not work properly: The space can only be filled with gaps. However, if you cut the vertices of an octahedron into regular hexagons, you can combine these solids seamlessly, so that the space is "tessellated" without any gaps (image to the left). There are – aside from the cube – only two such solids (the second one, the rhombic dodecahedron, is pictured at the top middle of the right-hand page).

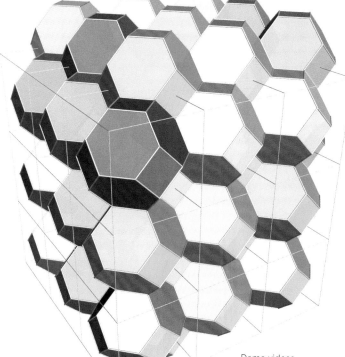

Demo videos
http://tethys.uni-ak.ac.at/cross-science/develop-dodeca.mp4
http://tethys.uni-ak.ac.at/cross-science/create-dodecahedron.mp4
http://tethys.uni-ak.ac.at/cross-science/soccerball.mp4

Transforming a cube

Take a cube and rotate its faces around the cube's twelve edges by the same degree α (left images).

Between $\alpha = 45°$ and $\alpha = 148.3°$, you will get to see some surprises.

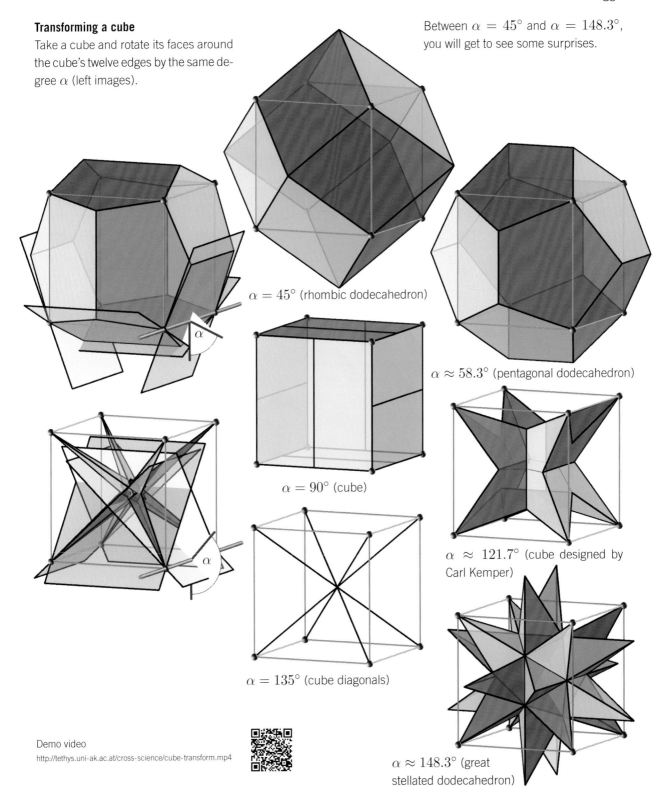

$\alpha = 45°$ (rhombic dodecahedron)

$\alpha \approx 58.3°$ (pentagonal dodecahedron)

$\alpha = 90°$ (cube)

$\alpha \approx 121.7°$ (cube designed by Carl Kemper)

$\alpha = 135°$ (cube diagonals)

$\alpha \approx 148.3°$ (great stellated dodecahedron)

Demo video
http://tethys.uni-ak.ac.at/cross-science/cube-transform.mp4

Spatial Tessellations

Only trivial with cuboids

If we tessellate with nothing but cubes or cuboids, we can obviously fill a space entirely. Other than that, this would not work with the other platonic solids: The angles of the edges do not fit together exactly, not even with dodecahedra, even though it *almost* looks like they would fit (the dodecahedra in the top left image are not completely regular).

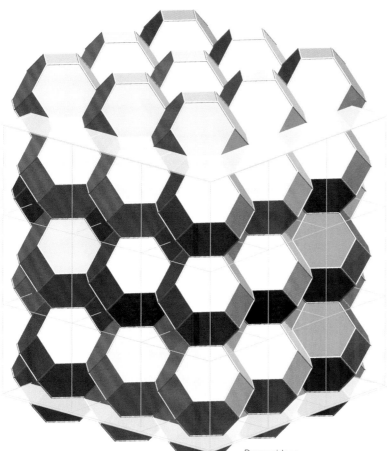

An Archimedean solid is suitable

For solids with two or three types of regular surfaces, we can determine that only one is suitable for stacking: a specially cropped octahedron – with six squares and eight regular hexagons (second image from the left and first video). The large image at the bottom left shows the tessellation in layers.

Another solid…

…can be found after a longer search – and nature has also "discovered" it with various stones, especially with garnets – hence the frequent labelling of such solids as granatoids instead of the otherwise common but cumbersome label rhombic dodecahedron, see the three images at the top right). There is not a single right angle (all corner angles are either $60°$ or $120°$), so that we often only mistakenly think that we can actually imagine such a solid. The second video shows the rotation of several such solids that have been joined together as well as some special views.

Demo videos
http://tethys.uni-ak.ac.at/cross-science/truncated-spacefiller.mp4
http://tethys.uni-ak.ac.at/cross-science/rhombic-spacefiller.mp4

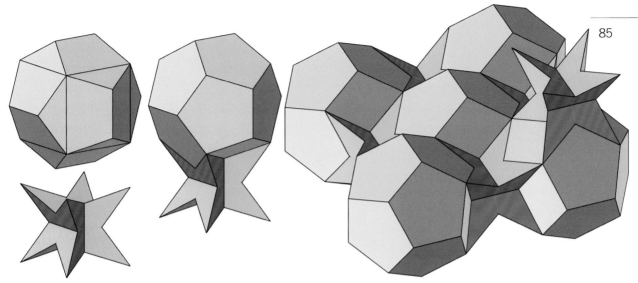

A complementary couple

As already mentioned the angles of do-decahedra make space-filling stacking im-possible. What fits exactly, however, are solids that look like an "egg and egg cup" (second image at the top left, first video): On p. 44 we could see that a plane can be tessellated without gaps if we use reg-ular and semi-regular pentagons. Hans Walser, who discovered this, also noted that a space can be tessellated without gaps if we use regular and semi-regular dodecahedra (second video).

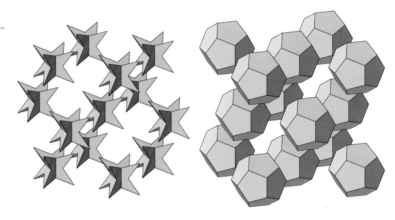

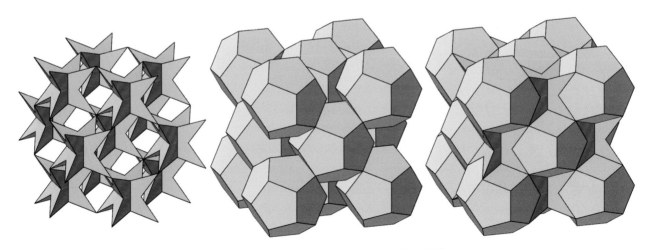

Demo videos
http://tethys.uni-ak.ac.at/cross-science/reg-and-semireg-dodecahedron.mp4
http://tethys.uni-ak.ac.at/cross-science/dodeca-spacefilling.mp4

Mobile or Not?

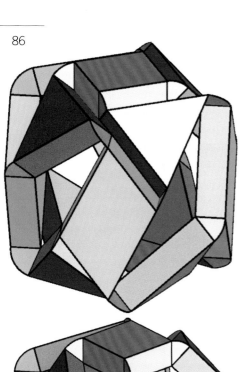

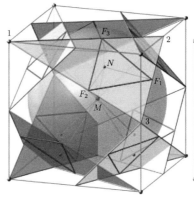

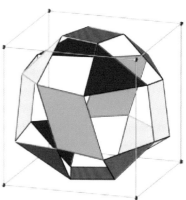

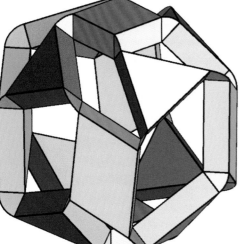

A surprisingly mobile mechanism

Take a cube and cut the vertices as shown in the top middle image (mind the orientation). It can be shown quite easily that every resulting blue triangle 123 touches the inscribed sphere of the cube in a point N. Then raise the perpendicular base points from N to the triangle sides and look at the resulting eight triangles.

Now we take the next steps: The eight red triangles also define two squares in the top and bottom surfaces of the cube as well as four congruent parallelograms (top right image). We then construct prisms of arbitrary but equal heights over the defined polygons. We connect the prisms with "spherical double hinges", for instance, in an arc-shaped manner, as in the top left image.

Actually, nothing should move

Such a multi-partite mechanism is usually rigid, i. e. it does not have any degree of freedom.

However, due to its special dimensions, this peculiar mechanism is astonishingly mobile (with "redundant constraints"). It can be moved without any problems until the edges of the prisms collide. In the image series to the left, the top and bottom images show two such extreme positions, and the middle image shows a general position.

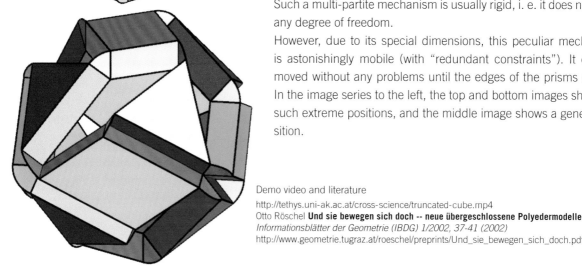

Demo video and literature
http://tethys.uni-ak.ac.at/cross-science/truncated-cube.mp4
Otto Röschel **Und sie bewegen sich doch -- neue übergeschlossene Polyedermodelle**
Informationsblätter der Geometrie (IBDG) 1/2002, 37-41 (2002)
http://www.geometrie.tugraz.at/roeschel/preprints/Und_sie_bewegen_sich_doch.pdf

How do you find such mechanisms with redundant constraints?

What has proven to be useful as a possible starting point for the construction of a series of polyhedron models with redundant constraints is the study of "equiform constrained motions", which are generated by superimposing a rotation and a similarity.

The "reduced Moebius mechanism" shown on the right was discovered through such constrained motions (the series at the far right shows a top view of this situation). The model at the bottom has been discovered in the same manner.

Demo video and image gallery
http://tethys.uni-ak.ac.at/cross-science/red-moebius.mp4
http://www.geometrie.tugraz.at/roeschel/bildgalerie.htm

Scutoids

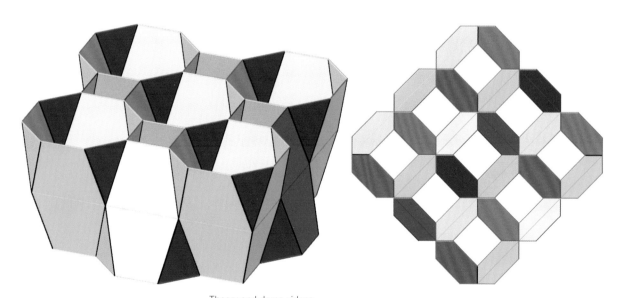

The name *scutoid*…

…has been coined by the authors of the paper cited below. They claim that "scutoids are a geometrical solution to three-dimensional packing of epithelia".

Connecting tilings

Take a number of regular polygons in a parallel plane and "connect them" – ideally with planes (first video) that then form truncated pyramids or general scutoids.

Connecting Archimedean tilings

If there are, as in the images on this page, only two types of polygons (here: squares and regular octagons) that, in addition, form an (Archimedean) tiling, and if we connect two such congruent tilings, we will soon get a beautiful constellation.

At the bottom right, we look at the three-dimensional scene from above, which reveals an interesting tiling that consists of squares and irregular symmetrical hexagons.

Theory and demo videos
https://www.nature.com/articles/s41467-018-05376-1.epdf
http://tethys.uni-ak.ac.at/cross-science/scutoid1.mp4
http://tethys.uni-ak.ac.at/cross-science/scutoid2.mp4

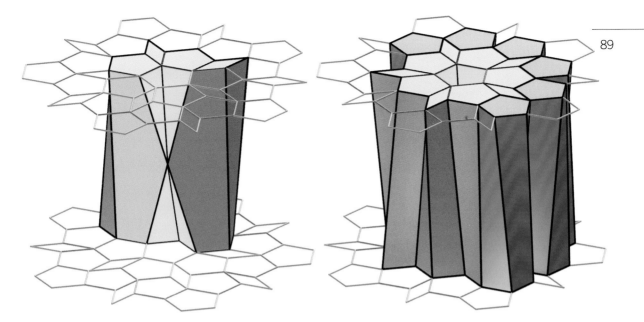

Aperiodic scutoid constructs

The two images on this page show the "linking" of two aperiodic tilings by means of scutoids. By definition, a scutoid has two parallel bounding polygons whose vertices are connected by either a Y-shaped link or a bent surface. As opposed to prisms, truncated cones and prisma-toids, a scutoid has at least one vertex between the two parallel bounding planes.

Doubly bent surfaces on the sides

In this specific case, the neighbouring scutoids also have sides that are not planar (hyberbolic paraboloids).

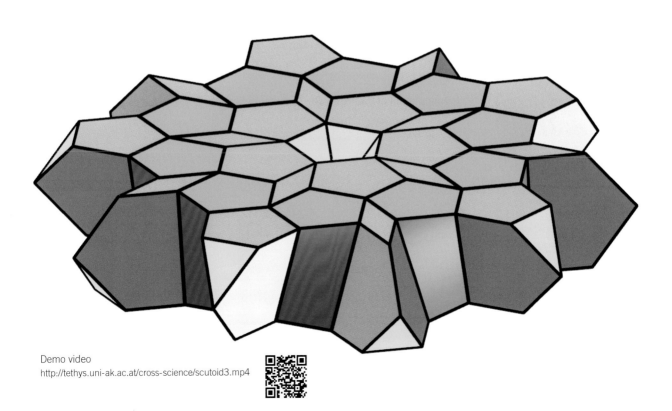

Demo video
http://tethys.uni-ak.ac.at/cross-science/scutoid3.mp4

Rubik's Cube

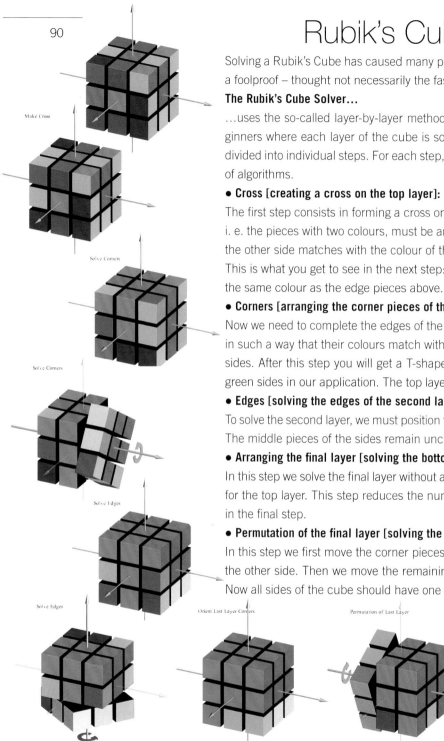

Make Cross

Solve Corners

Solve Corners

Solve Edges

Solve Edges

Solve Edges

Orient Last Layer Corners

Permutation of Last Layer

Permutation of Last Layer

Solving a Rubik's Cube has caused many people sleepless nights. Here you will find a foolproof – thought not necessarily the fastest – method:

The Rubik's Cube Solver…

…uses the so-called layer-by-layer method – a solution method often used by beginners where each layer of the cube is solved from top to bottom. The method is divided into individual steps. For each step, you only need to learn a limited number of algorithms.

● **Cross [creating a cross on the top layer]:**

The first step consists in forming a cross on the top (yellow) layer. The edge pieces, i. e. the pieces with two colours, must be arranged in such a way that the colour on the other side matches with the colour of the middle piece on each lateral surface. This is what you get to see in the next step: the green and blue middle pieces have the same colour as the edge pieces above.

● **Corners [arranging the corner pieces of the top layer]:**

Now we need to complete the edges of the top layer. The edge pieces are arranged in such a way that their colours match with the colours of the middle pieces on the sides. After this step you will get a T-shape, which can be seen on the brown and green sides in our application. The top layer is now complete.

● **Edges [solving the edges of the second layer]:**

To solve the second layer, we must position the four remaining edge pieces correctly. The middle pieces of the sides remain unchanged.

● **Arranging the final layer [solving the bottom without edge pieces]:**

In this step we solve the final layer without arranging the edge pieces the way we did for the top layer. This step reduces the number of possible cases to be considered in the final step.

● **Permutation of the final layer [solving the bottom corners and edges]:**

In this step we first move the corner pieces into their correct positions in relation to the other side. Then we move the remaining edge pieces into the correct position. Now all sides of the cube should have one colour.

Demo video and online explanation

http://tethys.uni-ak.ac.at/cross-science/rubiks-cube.mp4
https://www.youtube.com/watch?v=aDbZbr1Wa6E
https://de.wikipedia.org/wiki/Methoden_zum_L\%C3\%B6sen_des_Zauberw\%C3\%BCrfels

A Sphere with Varying Radius

The mega sphere by Chuck Hoberman…

…is a spherical plastic toy that expands and contracts as you pull it open or fold it down. The toy consists of six complete rings with Hoberman mechanisms (see literature cited below) that are all connected to one another.

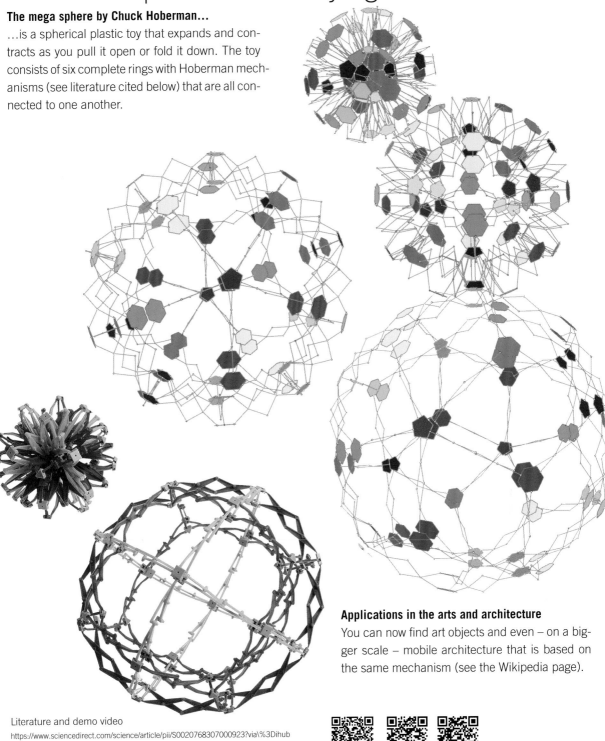

Applications in the arts and architecture

You can now find art objects and even – on a bigger scale – mobile architecture that is based on the same mechanism (see the Wikipedia page).

Literature and demo video

https://www.sciencedirect.com/science/article/pii/S0020768307000923?via\%3Dihub

https://en.wikipedia.org/wiki/Hoberman_mechanism

http://tethys.uni-ak.ac.at/cross-science/hoberman.mp4

Simple-Curved:
Developable!

Transforming and Developing in General Cases

Indentations as development

On page 28 we already discussed the oloid as a classic example of a solid that can be rolled on a plane, always touching the plane along a straight line (first video). From the indentations, we derived the development of the surface. Conversely, we can now cut out the development from paper and roll it together again. This simple method will yield a paper model of an oloid.

Some materials are malleable up to a certain extent

"Re-transforming" on a computer can lead, under certain circumstances when it is only done in an approximate manner, to minimal inaccuracies (consider the image series on this page as well as the second video cited below). The resulting surface should actually only have straight contour lines. In practice, however, we might get slightly bent contours, for instance, when we fold a thin sheet of metal, which is due to the compression or stretching of the material.

Demo videos
http://tethys.uni-ak.ac.at/cross-science/rolling-oloid.mp4
http://tethys.uni-ak.ac.at/cross-science/unrolling-surf.mp4

A paper model of a Moebius strip

Clearly the simplest example of the creation of a relatively complicated looking object is the twisting and then sticking together of a rectangular strip of paper. Due to the paper's stiffness, we get a strip that is dimensionally quite stable. Through this strip we can illustrate the most typical characteristics of a Moebius strip (namely that is non-oriented) quite well.

Demo videos
http://tethys.uni-ak.ac.at/cross-science/moebius-develop.mp4
http://tethys.uni-ak.ac.at/cross-science/moebius-with-colors.mp4
http://tethys.uni-ak.ac.at/cross-science/moebius-3-ribbons.mp4

A moebius strip is non-oriented

If we engrave a moebius strip with "scrolling text", then you will see the text once in forward and once in reverse direction (i.e. mirror-inverted).

Playing melodies forwards and backwards

If the engraving is a "crab canon" by Johann Sebastian Bach, then it can be played in both directions simultaneously. After the first voice, the musical notes will reverse. The melody will still fit harmonically with the first voice and could even be played simultaneously.

Demo videos
http://tethys.uni-ak.ac.at/cross-science/moebius-bach.mp4
http://tethys.uni-ak.ac.at/cross-science/moebius-loop-text.mp4

A moebius strip is developable

It thus carries a bundle of straight lines. Depending on where you cut the lines, you will get a section of the total surface area.

Moebius strips are typically known as relatively thin, twisted rectangular strips. However, if we cut the straight lines – as shown in the images on this page – with a much wider rectangle during developing, then we will get aesthetically pleasing, developable surfaces that can be identified as moebius strips when we draw in the narrower strip.

Demo video
http://tethys.uni-ak.ac.at/cross-science/moebius-variation.mp4

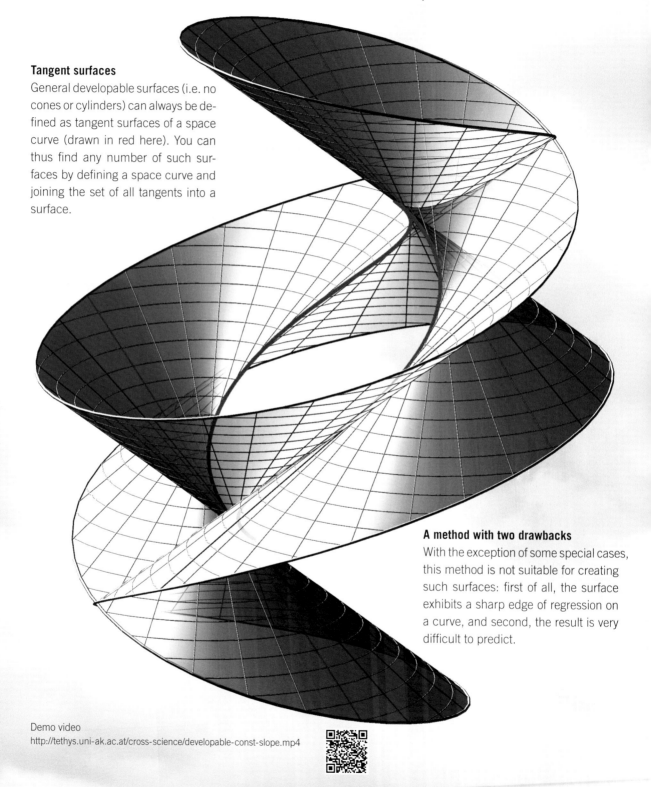

Tangent surfaces
General developable surfaces (i.e. no cones or cylinders) can always be defined as tangent surfaces of a space curve (drawn in red here). You can thus find any number of such surfaces by defining a space curve and joining the set of all tangents into a surface.

A method with two drawbacks
With the exception of some special cases, this method is not suitable for creating such surfaces: first of all, the surface exhibits a sharp edge of regression on a curve, and second, the result is very difficult to predict.

Demo video
http://tethys.uni-ak.ac.at/cross-science/developable-const-slope.mp4

The moving tripod

A regular space curve generally has a tangent and an circle of curvature in each point. The direction to the centre of this circle is known as principal normal. The tangent and the principal normal define the so-called binormal, which is perpendicular to the carrier plane of the circle. The tangent, the principal normal, and the binormal, in turn, define the so-called moving tripod, and thus three planes (drawn in green, red, and yellow).

We know that planes always envelop a developable surface in the course of a motion, and we apply this knowledge to the tripod: The green plane envelops the tangent surface, which, as we know, is suitable only to a limited extent for the targeted construction of developable surfaces. The yellow plane (perpendicular to the tangent) is useless for our purposes because it sweeps across a barely predictable surface. The red plane, on the other hand – also known as rectifying plane – generates a developable surface that contains the space curve as a geodesic line. This "rectifying torus" is particularly suitable for the design of our surfaces.

Creating ribbons

During the development of a rectifying surface, the corresponding space curve turns into a straight line. If we now cut the generatrices through straight lines in a parallel direction, we get ribbons of constant width. Problems arise, however, from the inflection points of the curve, because the tripod is not defined there. Such points must therefore be circumvented through suitable neighbouring points.

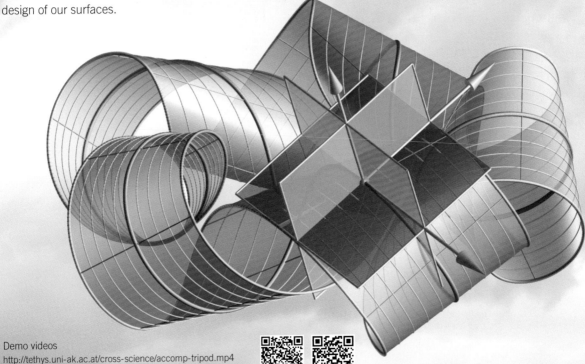

Demo videos
http://tethys.uni-ak.ac.at/cross-science/accomp-tripod.mp4
http://tethys.uni-ak.ac.at/cross-science/torsion.mp4

Mobile Though Theoretically Not Possible

Developable connecting surfaces

There are n larger and smaller circles that are uniformly distributed on meridian planes. They define $2n$ congruent developable connecting surfaces. Max Klammer examined the creation of such partial surfaces, which resembles the creation of truncated cones. He connected them in the manner suggested here – first with "toothed folds", later also with paper clips and staples. Klammer discovered that the entire object – by now known as "Klammer torus" – can be interpreted as a machine that can revolve as long as it remains "a little flexible" and only parts of the surfaces are materialised.

Demo video
http://tethys.uni-ak.ac.at/cross-science/klammer-torus.mp4

Collar Surfaces

Carving of curves

Take a rectangular strip of drawing paper, carve a curve c into the paper and shape the paper along the carved line carefully into a cylinder of revolution. You will get a series of "collar surfaces". In the images above, a catenary has been chosen for c. Other options are sine waves (images below) and parabola. However, the carved curve cannot have any inflection points. Since nothing changes about the surface's metrics (otherwise the paper strip would tear), you will have found a class of developable and thus simple-curved surfaces.

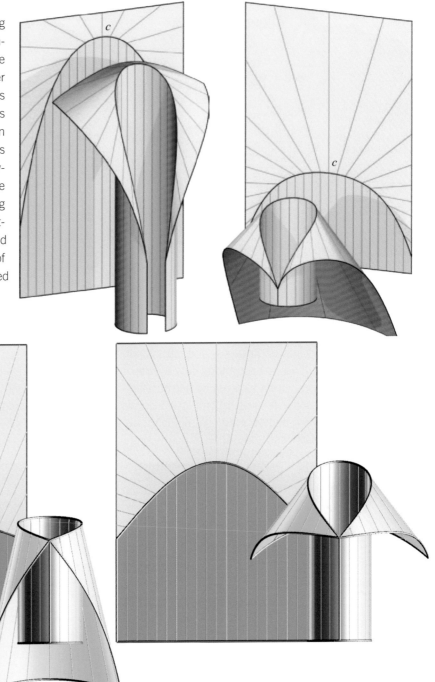

Demo videos
http://tethys.uni-ak.ac.at/cross-science/collar-surface1.mp4
http://tethys.uni-ak.ac.at/cross-science/collar-surface2.mp4

Surfaces with Locally Constant Slope

Sand dunes are locally simple-curved surfaces

Let us look at the images on the right-hand page (sand dunes in Namibia). If we let sand trickle onto a certain spot like in an hourglass, it will form a cone with constant slope. If we do the same in all points of a spatial curve (right-hand page in the middle, image on the right), we get the enveloping surfaces Γ_1 and Γ_2 of all correspond-

ing cones, which then also have a constant slope. This "dune surface" has a trace curve on the base plane. On the other hand – starting from the trace curve – we can reconstruct the dune. For the fun of it, this has been illustrated with the example of the Earth, which demonstrates that dune surfaces have a constant slope only locally.

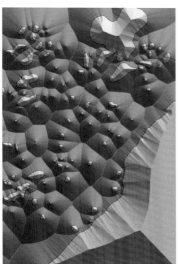

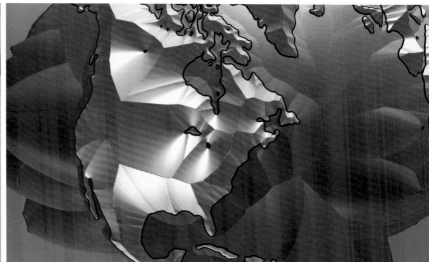

Demo videos
http://tethys.uni-ak.ac.at/cross-science/hourglass.mp4
http://tethys.uni-ak.ac.at/cross-science/pouring-gravel-heap.mp4

Circulation and Torsion

Circulation of a "rhombic chain"

Rhombi have four equally long sides and are normally planar. However, they can be bent along a diagonale and labelled as "skew rhombi".

The following idea is not trivial: let us distribute points uniformaly on a torus's circle of latitude. They act as vertices of the rhombi on the diagonale that is not bent. Their perpendicular bisector planes cut through the torus's meridian circles, on which – provided that the side lenghts are given – you look for the remaining points of the rhombi. You have thus already found a circulating rhombi chain.

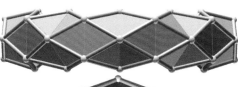

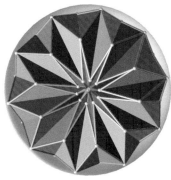

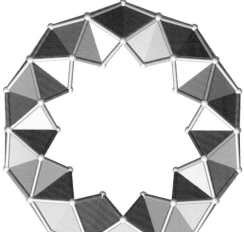

Now the side lengths of the individual rhombi can be chosen such that the construct is closed after multiple applications. Since several parameters can be varied, you will get an astonishing multitude of such closed constructs, which can all be twisted like a kaleidoscope.

Demo videos
http://tethys.uni-ak.ac.at/cross-science/diamond-mesh-torus.mp4
http://tethys.uni-ak.ac.at/cross-science/diamond-mesh-torus2.mp4

Twisted rings

Let us take a ring with square cross-section and twist the cross-section with a widening centre angle. The result is bounded by surfaces that can be described as generalised Moebius strips.

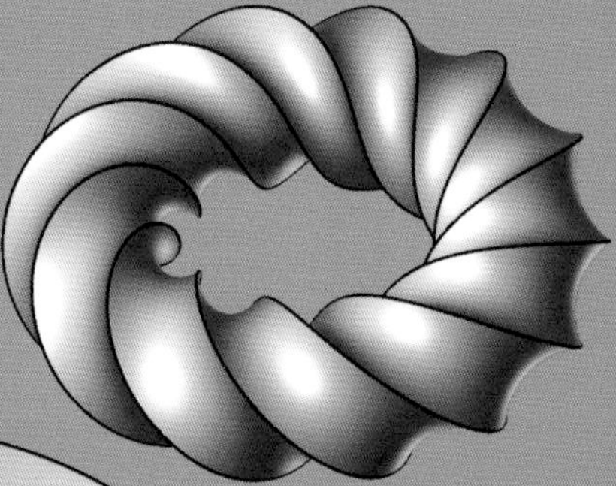

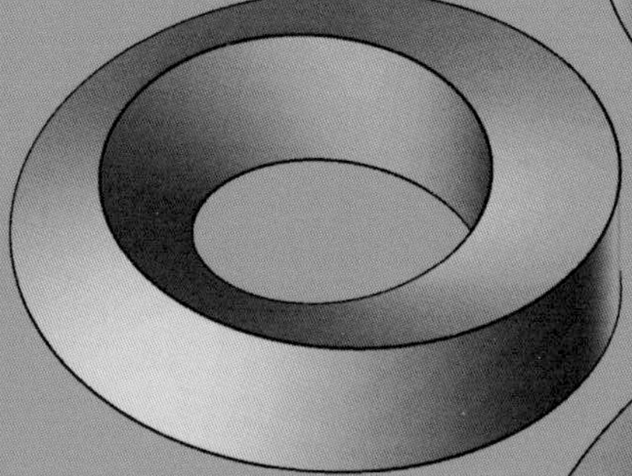

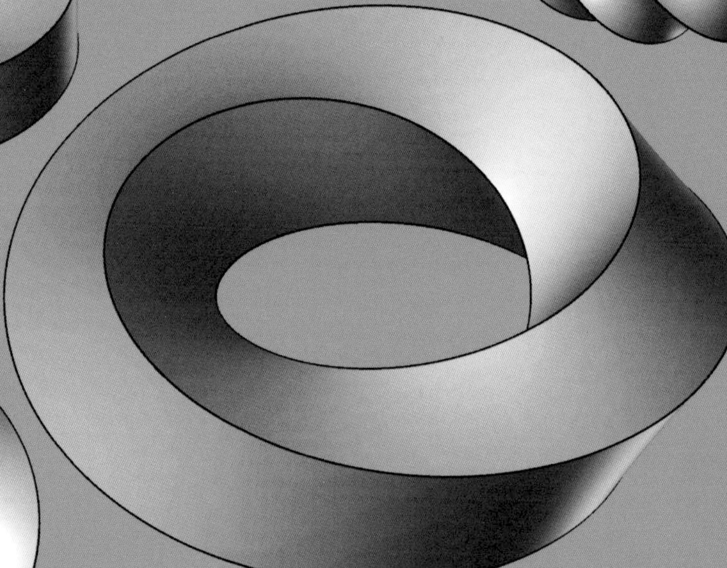

Double-curved

Even if the aforementioned boundary surfaces are "almost" developable, they still have slightly curved contours, which suggests that they are not simple-curved and thus developable. It is, therefore, not possible to create a paper model of the rings pictured here.

Demo video
http://tethys.uni-ak.ac.at/cross-science/moebius-generalization.mp4

Double-Curved: The Norm

Football Variations

Criteria for a football

The following international rules hold: A football must be spherical, made of leather or some other suitable material, have a circumference of $68 - 70$ cm, a mass of $410 - 450$ g, and an inside pressure of 600–1100 g/cm^2. These are the basic facts.

The sphere is double-curved and thus not developable

What is especially important is the exactly spherical shape of the football – otherwise the ball cannot roll evenly in all directions. In order to produce such a ball, the individual pieces, which are usually made of leather, are stitched together, though a plastic "bladder" is placed inside the still spongy shape.

The shell's deformability is crucial

Filling this inner bladder with air will turn the ball into a sphere. The leather pieces that have been stitched together are thus pushed out. The more evenly the pieces have been distributed, the better this will work. As is well know, it is impossible to combine only congruent plane polygons (the image below shows a good "triangulation" but not all the triangles are of the same size).

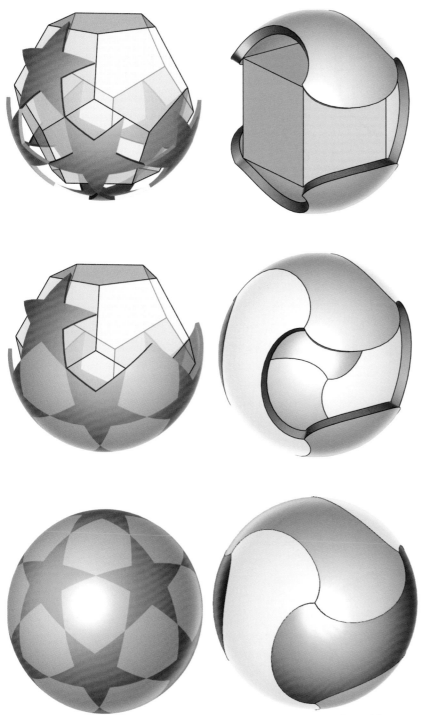

Platonic base bodies

The most obvious base structure is probably a pentagonal dodecahedron. The classic shape of a football (photograph at the top of the left-hand page) is derived via a dodecahedron from an Archimedean solid with twelve pentagons and 20 hexagons. From time to time new football designs are developed.

Twelve congruent stars

The image series on the very left with the twelve yellow stars shows the football design for the Champions League final held in Munich in 2012. You can clearly recognise the dodecahedron underlying its layout.

Only six congruent pieces

The second image series shows a design based on a cube whose shell consists of only six congruent pieces. For this design the indentations and protrusions must be congruent. At the same time, the eight cube vertices are supposed to form regular "vortices" (bottom right image). In the concrete case, we are dealing with true spatial curves that are generated as intersections of elliptic cylinders with the sphere. Yet, the material of which the individual lobes are made must be very elastic, because there is a strong double curvature.

Demo videos
http://tethys.uni-ak.ac.at/cross-science/fussball1.mp4
http://tethys.uni-ak.ac.at/cross-science/fussball2.mp4

Gardeners on Curved Surfaces

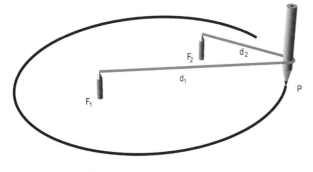

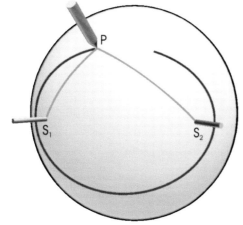

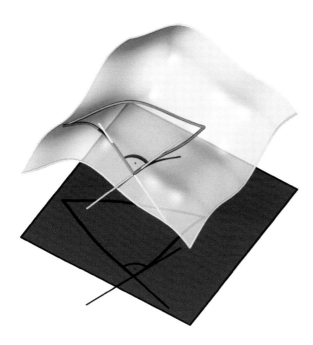

A simple ellipse construction

An ellipse is distinguished by the fact that the sum of the distances d_1 and d_2 to two fixed points F_1 und F_2 is constant. This characteristic enables a method of drawing known in German as "Gärtnerkonstruktion" (gardener construction): Fix the two points and guide a rope or string of constant length around as pictured in the image to the left.

The same construction on a sphere

This construction can be generalised on the sphere (image below). The two rope sections are obviously curved in this case. When the rope is straightened, the two sections will cover the shortest distance between the fixed points and the curve point and thus run along two great circles.

Generalisation on any surface?

Now you might wonder if such a gardener construction could be expanded to all surfaces. However, you might have to hold the rope on the surface as if through a magnet to make sure that it will not detach and form a straight-lined "bridge".

Geodesic lines

Let us look for the shortest connection between two points of a surface. From geometry we know that this must be a geodesic line, and we already know an example of such a line: On the sphere, great circles are geodesic lines.

Per definition, such a line is characterised by the fact that, in all its points, the corresponding osculating plane contains the surface normal. This may sound complicated (and does generally require complex calculations), but one can prove that the curve that we are looking for must fulfil this requirement. In the case of the sphere, geodesic lines are plane curves, and the plane does, in fact, intersect the sphere at a right angle because it passes through the sphere centre.

A "right triangle" has been drawn on a surface in this manner in the image to the right. All three sides are geodesic lines on the surface.

Construction on elliptically curved surfaces

The images on this page show "ellipses" on an ellipsoid (top), on a one-sheeted hyperboloid (middle) and on a torus (image at the bottom). Ellipsoids are always elliptically curved. That is why the construction always works perfectly, also physically: The string will automatically lie on the surface.

Critical positions on hyperbolically curved surfaces

With a one-sheeted hyperboloid (which is hyperbolically curved in all its points), the string will occasionally try to take a "shortcut" by lifting itself in a straight line from one point to another on the surface (in the image, for instance, from the pecil tip to the fixed point on the back side). If both fixed points lie either at the top or the bottom, then the problem is partially resolved.

Gardener construction on a torus

A torus is hyperbolically curved inside and elliptically curved out-side. In this case, a physical construction is only locally possible, and the whole thing should instead be seen "from a philosophical perspective": Imagine a two-dimensional organism that lives on a torus and cannot leave it. Gravity always acts orthogonally to the surface. The "flatlander" on the torus will no doubt believe that they are drawing an ellipse with their string construction.

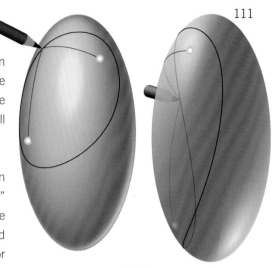

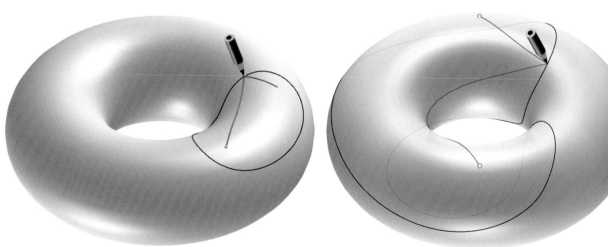

Demo videos
http://tethys.uni-ak.ac.at/cross-science/gardener-on-quadric.mp4
http://tethys.uni-ak.ac.at/cross-science/gardener-on-torus.mp4

The Tassilo Chalice is a chalice that is kept at Kremsmünster Abbey in Austria. It was probably donated by the Bavarian Duke Tassilo and his wife Luitpirga around 780 AD (see foonote for more details). Geometrically speaking, the chalice is made of an ellipsoid of revolution (upper part), a surface of revolution that resembles the outer shell of a torus (middle part), and an approximation of a hyperboloid (bottom part, pedestal).

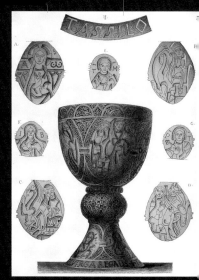

Five decorated ovals at the top and four at the bottom…

…are evenly distributed across the double surfaces. This requires considerable geometrical knowledge.

First theory

Was the upper part of the chalice first approximated by the artist through a cylinder of revolution and the lower part through the frustum of a cone of revolution? The corresponding developments would then be a rectangular strip and an annulus sector. It would not be difficult to draw the ovals on these and then "roll" metal leaves onto it. However, in practice this procedure would not work very well, though incisions in the metal would give you some wiggle room.

Information about the Tassilo Chalice
Historical overview, photographs, ornaments
https://en.wikipedia.org/wiki/Tassilo_Chalice
https://mittelalter.fandom.com/de/wiki/Tassilokelch

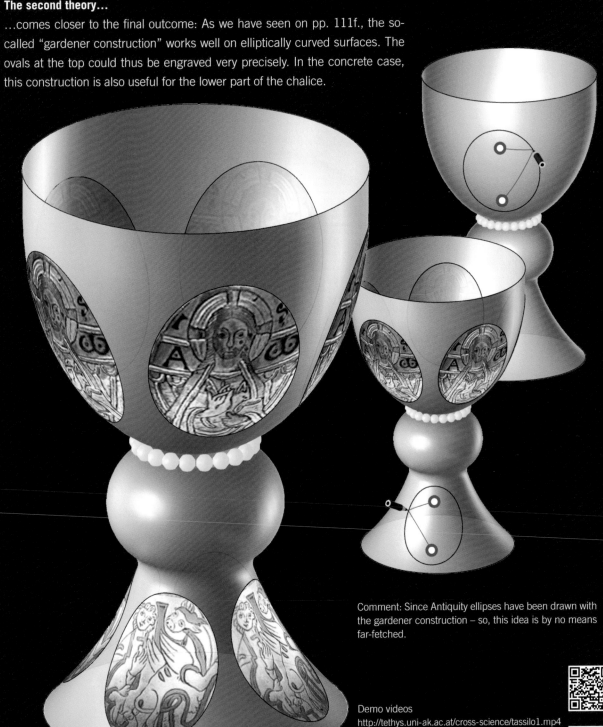

The second theory...

...comes closer to the final outcome: As we have seen on pp. 111f., the so-called "gardener construction" works well on elliptically curved surfaces. The ovals at the top could thus be engraved very precisely. In the concrete case, this construction is also useful for the lower part of the chalice.

Comment: Since Antiquity ellipses have been drawn with the gardener construction – so, this idea is by no means far-fetched.

Demo videos
http://tethys.uni-ak.ac.at/cross-science/tassilo1.mp4
http://tethys.uni-ak.ac.at/cross-science/tassilo2.mp4

Animal Horns

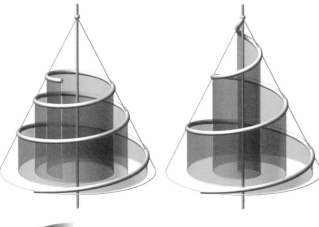

Two not so dissimilar spatial spirals

Spirals are commonly understood to be "a kind of screw". In geometry, however, we distinguish between different types of spirals. If, for instance, a shape is rotated around its axis and simultaneously moved in proportion to the rotation across the axis, then we speak of a screw. If a shape axially shrinks in proportion to the rotation angle, then we call it a helix. When the shrinking is exponential in relation to a fixed centre on the axis, then we are dealing with a classical spiral. The difference between classical and helix spirals can be seen in the top left image: While the former never reaches the spiral centre, helix spirals reach a point on the axis with no problem.

Helix

An analysis of the famous painting "The Tower of Babel" by Peter Breughel the Elder (image in the middle) reveals, for instance, a very precise helix. If the painter had created a higher tower, it would have taken only a few more rotations for the building to reach its tip.

Helical surfaces

If a circle in a meridian plane is subjected to a helix spiral, as in the image below, we get a surface that is strikingly reminiscent of animal horns. As shown by the images on the right-hand page, we can create differently shaped animal horns through variation of the initial curve. This gives us a better understanding of how such horns grow in nature.

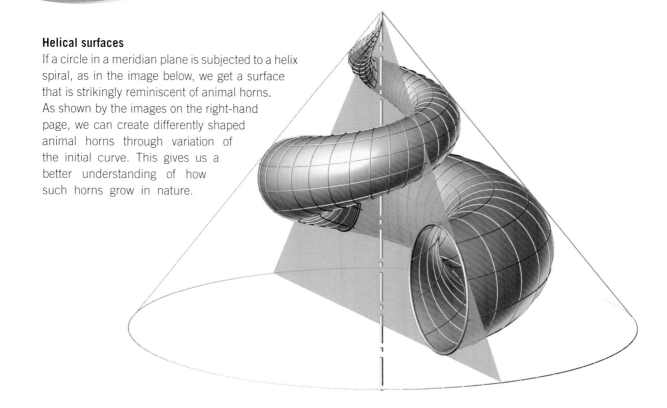

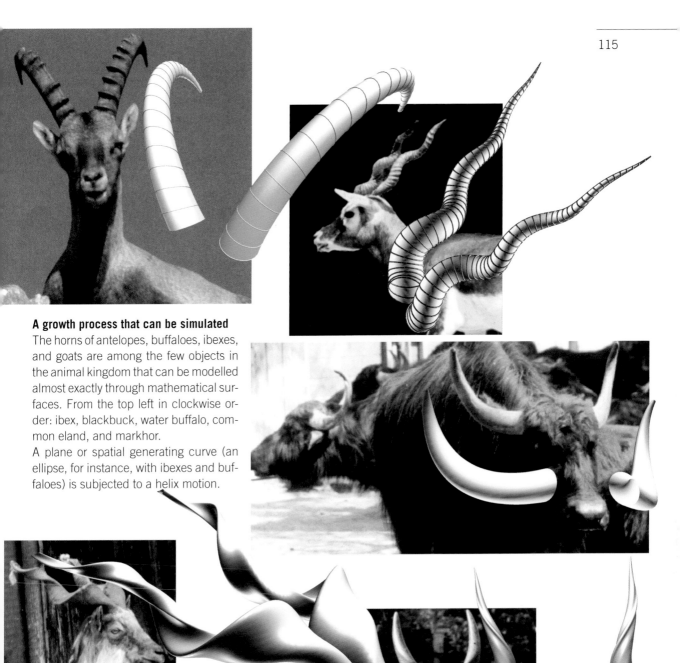

A growth process that can be simulated

The horns of antelopes, buffaloes, ibexes, and goats are among the few objects in the animal kingdom that can be modelled almost exactly through mathematical surfaces. From the top left in clockwise order: ibex, blackbuck, water buffalo, common eland, and markhor.

A plane or spatial generating curve (an ellipse, for instance, with ibexes and buffaloes) is subjected to a helix motion.

Demo videos
http://tethys.uni-ak.ac.at/cross-science/markhor.mp4
http://tethys.uni-ak.ac.at/cross-science/eland.mp4

Exponential Growth

Snail shells

The connection between geometry and biology is probably most beautifully visible with the calcium shells of snails, mussels, and nautiluses. In a top view, the growth lines appear as logarithmic spirals, that is, as curves that intersect a radial pencil of rays at a constant angle. Such curves emerge when the radial distance of a point is increased by a certain percentage and the point is simultaneously rotated around the centre by a *proportional* angle. This is precisely the kind of growth that can be observed with the aforementioned animals.

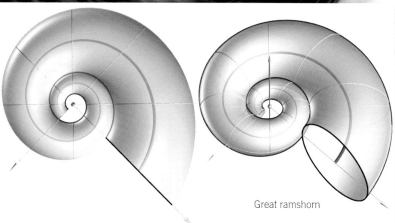

Great ramshorn

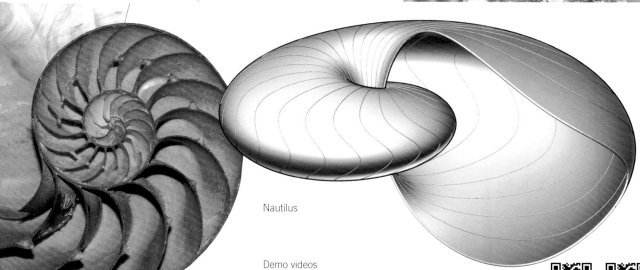

Nautilus

Demo videos
http://tethys.uni-ak.ac.at/cross-science/ramshorn.mp4
http://tethys.uni-ak.ac.at/cross-science/nautilus.mp4

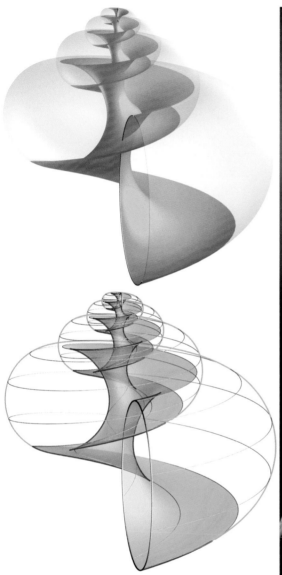

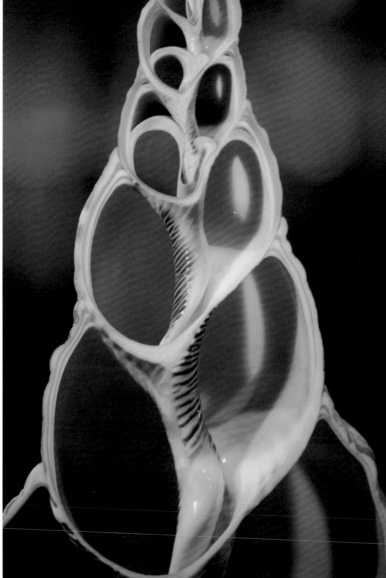

Fascinating inner life

As straightforward as snails shells might appear on the outside: The inside is far more complex and requires great imagination, especially, because there are almost always self-intersections (an exception are the shells of great ramshorns – pictured to the left and in the middle – where, remarkably, self-contact occurs).

Demo video
http://tethys.uni-ak.ac.at/cross-science/snail-shell-section.mp4

Radial Symmetry

Symmetry…

…obviously plays a very significant role in nature, both with plants and animals, due to a variety of reasons. The deer, pictured at the bottom right of the left-hand page, would not be able to effortlessly balance its massive antlers and would need to exert greater force to move its head with ease if its antlers were not symmetrical. Vertebrates, but also most other phyla in the animal kingdom, possess a single symmetry plane: They are bilaterally symmetrical and thus have a "front side" – the direction of movement.

Jellyfish (top right on the left-hand page) and other cnidarians are radially symmetrical. These animals also move in all directions. On the left-hand page, you can see two more examples of radial symmetry: Acetabularia algae, also known as mermaid's wineglass, at the top left and a fly agaric below.

Radial symmetry more common in the plant kingdom

As opposed to the animal kingdom, the plant kingdom seems to feature bilateral symmetry less frequently. Manifold symmetries, such as the fivefold symmetry of bellflowers, can be found more often. Biologists believe they know the reason for this: Flower calyces are supposed to attract insects that pollinate the flowers. If a calyx looks almost the same from all sides, it forms a landing platform that insects can approach from all directions. This has proven to be very beneficial, even if it means that pollen must be applied on all sides of the calyx (producing pollen costs a lot of energy).

Geometrically speaking…

…manifold symmetries usually involve surfaces of revolution, that is, surfaces that are generated by the rotation of a meridian around a fixed axis. The bellflower with its fivefold symmetry, for instance, has a surface of revolution as a chalice shape (see video).

Demo videos
http://tethys.uni-ak.ac.at/cross-science/blossom-variations.mp4
http://tethys.uni-ak.ac.at/cross-science/medusa-moving.mp4

Classical Surface Types

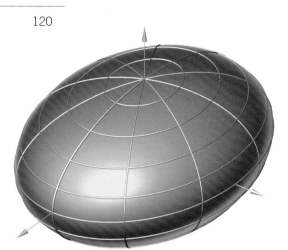

The simplest surfaces

Aside from planes, surfaces of the second order are the simplest algebraic surfaces. In nature, the developable variants (cylinders and cones) and of course spheres play the most important roles. In differential geometry they serve as prototypes for the local behaviour of all surfaces. The image to the left shows a triaxial ellipsoid.

Surfaces of revolution

Among the surfaces of the second order, we will also find surfaces with rotational symmetry. Saturn, for instance, has, more than all the other planets, the shape of a flattened ellipsoid of revolution.

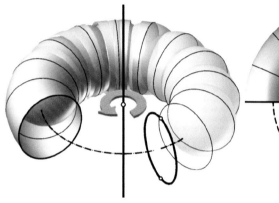

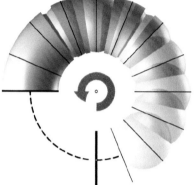

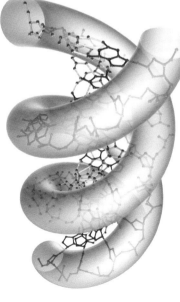

In geometry the torus that is generated by the rotation of a sphere or a circle in a meridian plane, is a suitable prototype for the approximation of any surface of revolution (images in the middle).

Pipe surfaces

In nature, surfaces that are generated by a moving sphere often play an important role. At the bottom left, you can see a screw surface where the sphere has been subjected to a screwing motion. The image to the right of it shows a double helix as the structure of molecules in the chromosomes of a cell's nucleus. The DNA strands spiral around an imaginary axis.

"Surface hybrids"

Surfaces of the second order can be surfaces of revolution, even if, at first, they may not look like it. The paraboloids pictured to the right, on the other hand, are always translation surfaces (p. 130). The one on the left, however, could also be a surface of revolution. The paraboloid on the right is generated in a twofold way by moving a straight line and is thus also a "ruled surface".

Surface of revolution and surface of translation

The surface below – a surface of revolution that is generated by the rotation of a sine wave around its axis – can be interpreted as a midsurface of two coaxial helices (left- or right-handed, p. 130) that have been shifted by half a pitch. So, it is created in a nontrivial manner through the translation of one helix along another (video 1).

Helical surfaces, ruled surfaces, and translation surfaces

The helicoid (image below) is, first of all, a helical surface with half a parameter, then a ruled surface because it is generated by a straight line that hits the axis vertically, and finally also a translation surface in infinitely many ways:

All it takes is a helix with a parallel axis and half a parameter that intersects the axis of the surface and translates it along an identical helix (video 2). This also explains why you can always find an orthographic projection of the surface in which such a nontrivial helix forms the surface's contour.

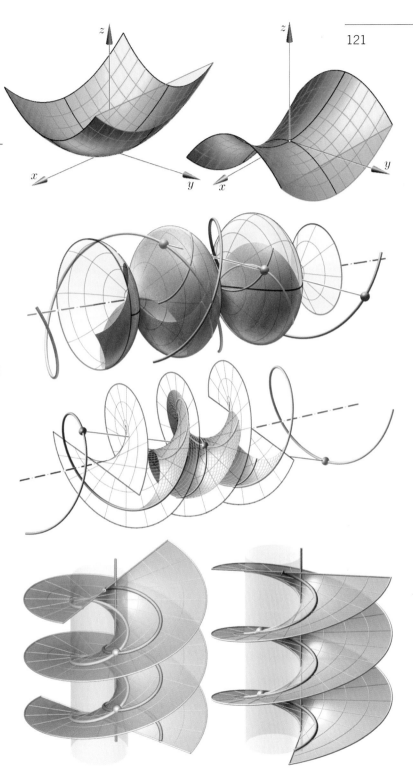

Demo videos
http://tethys.uni-ak.ac.at/cross-science/rotating-sine-translation.mp4
http://tethys.uni-ak.ac.at/cross-science/helicoid-translation.mp4

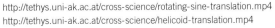

A Peculiar Type of Constrained Motion

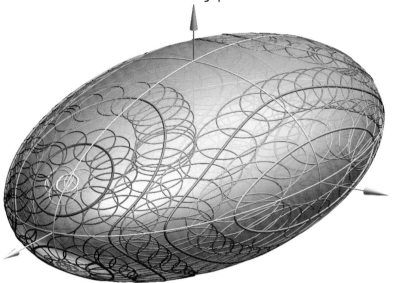

A remarkable problem

Oloid motions (p. 28) were already quite difficult to explain. Yet, there is another interesting motion that was not fully explained until the year 2020. The following problem had to be resolved: Take a triaxial ellipsoid and intersect it with an arbitrary plane. The result will always be an ellipse. Since a plane is defined by three points, this can be done in ∞^3 ways. However, the shape of an ellipse is already clearly defined when the lengths of the major axis and the minor axis are given. So, there are only ∞^2 different ellipses. As a result, an ellipsoid must contain infinitely many *congruent* ellipses. Is it maybe possible then to move a given ellipse on an ellipsoid, and, if so: What kind of motion is this?

A continuous motion

The image to the left illustrates how ellipses of varying "widths" can be moved on a triaxial ellipsoid. The exact solution to this problem is – if you want to have a formula – very demanding. You can get an approximate approach to a solution through the considerations described on the right-hand page.

It is also worth mentioning that, by following very similar considerations, you can also move ellipses on triaxial hyperboloids (bottom-left image) and parabola on square cones (bottom-centre) or on hyperboloids (bottom-right).

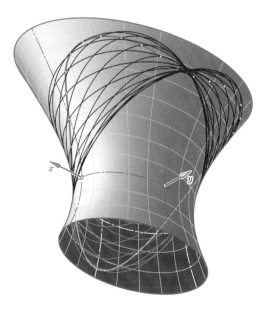

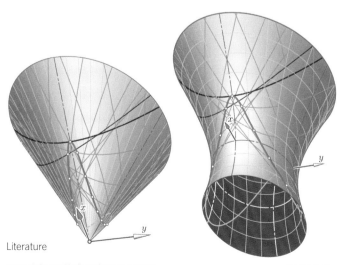

Literature

Hellmuth Stachel **Moving ellipses on quadrics**
*G -- Slovak Journal for Geometry and Graphics 17
no. 33, 29 -{42 (2020) (ISSN 1336-524X))* 2020
https://www.geometrie.tuwien.ac.at/stachel/190_G_ellipses.pdf

First some indispensable theoretical input

Triaxial ellipsoids span different widths in each main direction. Let us now look at an arbitrary point on such a surface. If we intersect an ellipsoid with a plane parallel to this tangential plane, we get ellipses that are similar to each other: The aspect ratio of the two axis lengths is the same for all of these ellipses.

Then the transition to standard algorithms

The only points that are of interest here are those that yield an aspect ratio of w_0 for the ellipse to be moved. To find these points quickly, we must parametrise the ellipsoid (parameters u and v), determine the aspect ratio w of the axis lengths for a sufficiently large number of points $P(u, v)$, and thus get a function graph (blue in the images) in a (u, v, w) coordinate system. The contour line of the function graph at a height of w_0 will lead to the corresponding points on the ellipsoid. Such algorithms are standardised and fast.

Now, among the infinitely many parallel cross sections, we just have to find the one that is the right size. For this, it is enough to determine e. g. the length of the ellipse's major axis and then find, once again, the correct solutions (two or zero solutions) with a standardised algorithm.

We have thus found the locations of the ellipses in a continous function and can then e. g. visualise the trajectory of the vertices.

Interaction of exact and empirical solutions

Even among seasoned theoreticians, it is now common practice to draw on such visualisations in order to estimate if it is worth getting a perfect grip on a problem with complex calculations.

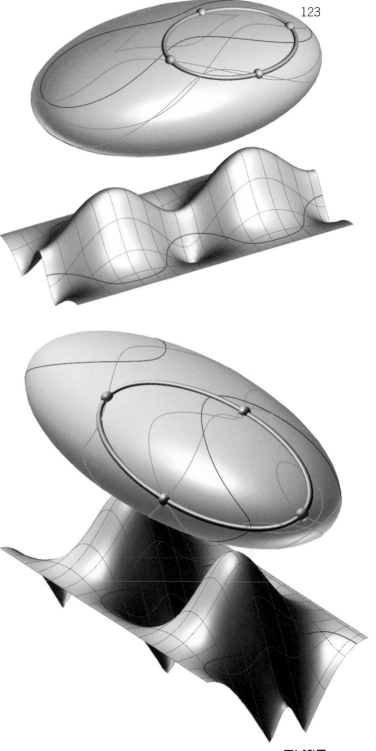

Demo video
http://tethys.uni-ak.ac.at/cross-science/move-ellipses.mp4

Special Nets on Cyclides

Can a curved surface be covered with equilateral triangles?

If you want to cover a cylinder of revolution without gaps using equilateral triangles, it can be achieved when the circumference fulfils certain criteria: You must develop the cylinder and inscribe a pattern of equilateral triangles. During development, the triangle sides that are not axially parallel and axially normal turn into sections of helixes.

On double-curved surfaces of revolution, we try this by means of so-called loxodromes, or rhumb lines, which run through the net of meridian curves at a constant angle. We already know loxodromes on spheres: They are spirals that transition into logarithmic spirals during stereographic projection from the poles (p. 69).

Loxodromes on the tori

You can usually only find loxodromes on tori (top left image) by solving differential equations. Among them, there are also the non-trivial *Villarceau circles* (clearly visible in the middle image). An especially beautiful net of triangles that only have $60°$ angles can be found if we seek for these circles the loxodromes with a course angle that deviates from the angle of Villarceau circles by $\pm 60°$ (images at the bottom).

Loxodromes on Dupine cyclides

Dupine cyclides are closely related to tori. They are generated through the inversion of the torus on a sphere that is centred around a point lying in the reference circle plane (top middle image on the right-hand page). The inversion is angle-, circle-, and sphere-preserving – precisely the characteristics that we need to transform the triangle net without damage (images on the right-hand page).

Demo video
http://tethys.uni-ak.ac.at/cross-science/torus-net.mp4

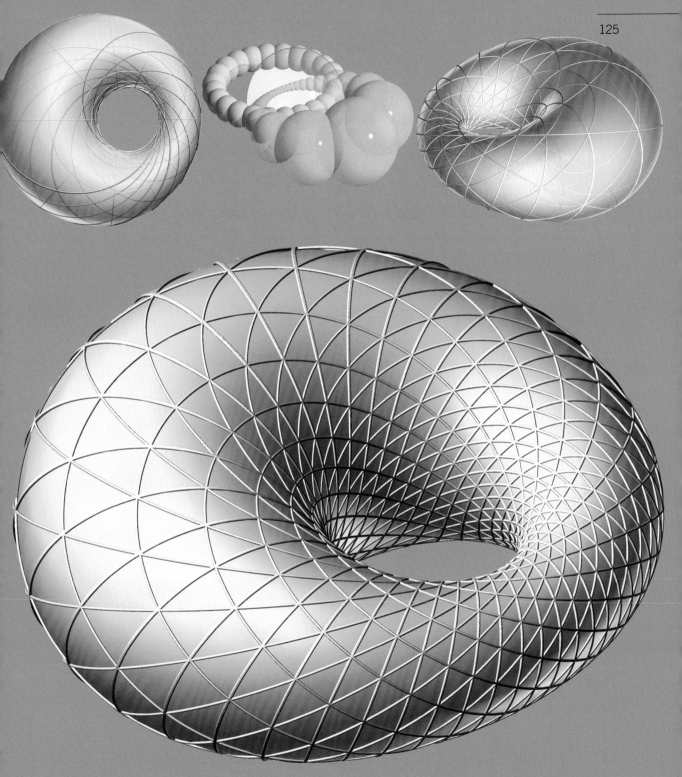

Demo video
http://tethys.uni-ak.ac.at/cross-science/cyclid-net.mp4

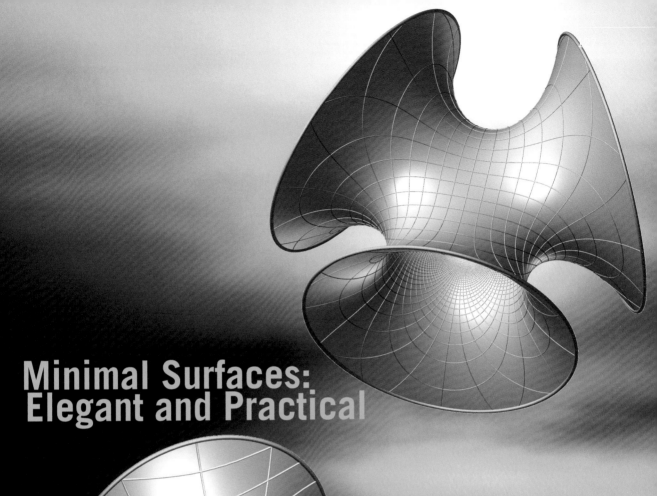

Minimal Surfaces:
Elegant and Practical

The Smallest Possible Surfaces

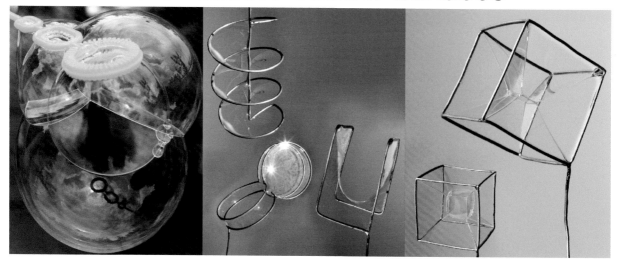

Spheres

Spheres are truly remarkable surfaces. Any plane intersecting a sphere would cut out a circle. If they pass through the sphere centre, the result is a great circle, though the plane is then a symmetry plane of the sphere – the sphere thus possesses a two-parametric bundle of symmetry planes and can be approached as a surface of revolution in just as many ways. All great circles are simultaneously geodesic lines (p. 110). Moreover, sphere are surfaces that envelope a given liquid volume with the smallest possible surface area. If you would like to create, for instance, a cube that fits exactly one litre of water, then the surface area of this cube will be more than 20% larger than that of a sphere holding the same amount of water. Viewed in this light, spheres could be described as "closed minimal surfaces". Small water drops (p. 126) are thus spherical, because the surface tension pulls together the surface surrounding the water to a minimum.

Minimal surfaces that are not closed

It gets more complicated if we would like to fit a surface area into a given structure in such a way that the surface becomes minimal. In the top middle and top right images (photographs by Katharina Rittenschober), you can see how such surfaces form quickly after dipping a special wire construct into soap. However, these surfaces are usually only stable for a short amount of time and then "burst".

Mathematical criteria

From an exactly mathematical point of view, this problem is generally difficult to resolve. It turns out that such surfaces require the main curvatures *in all points* to be inversely equal. All these surfaces are thus – unlike the sphere – hyperbolically curved. These *true minimal surfaces* are thus in all points, locally seen, bent like an equilateral hyperbolic paraboloid.

Classical minimal surfaces...

...include the catenoid (right-hand page, orange surface at the bottom) and the helicoid (right-hand page, yellow surface at the bottom). The twisting of these two surfaces into each other will be discussed on p. 131.

Another famous example is Costa's minimal surface (right-hand page, top images), which was first described in 1983. Topologically speaking, it can be defined as a thrice-punctured torus.

The trinoid (p. 127) can also be added to this list.

Demo videos
http://tethys.uni-ak.ac.at/cross-science/right-helicoid.mp4
http://tethys.uni-ak.ac.at/cross-science/costa-surface.mp4

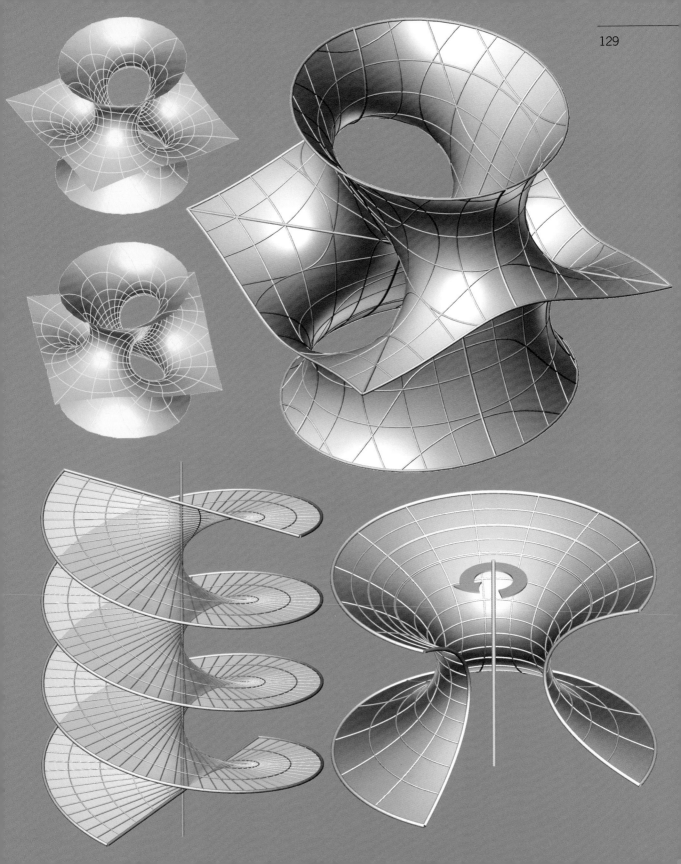

Associate Minimal Surfaces

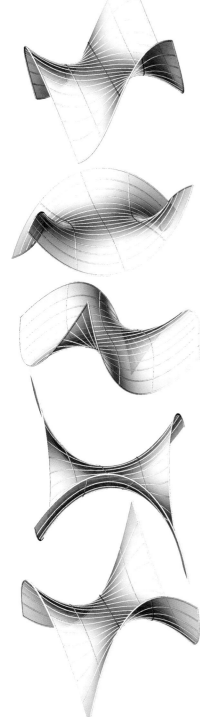

Isotropic curves

Can you imagine a spatial curve that has a length of zero? Obviously not – and yet, such curves exist in six-dimensional space (three real and three imaginary coordinates each), which we cannot possibly imagine. These curves are known as isotropic curves. They can be algebraic curves, helices, spirals, etc. You will soon understand why these curves have a practical value with visible results.

Translation surfaces

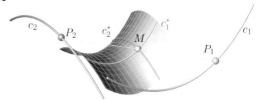

The image above shows how we can generate so-called translation surfaces: Take two curves c_1 and c_2, choose an arbitrary point P_1 and P_2 on each curve, and look at the centre M of the line spanning the distance between the two points. If you do this for all points of c_1 and c_2, you will get a surface that is generated by the translation of a (green) curve c_1^* along a (red) curve c_2^*.

Minimal surfaces as translation surfaces

Now the following theorem applies: If c_1 is an isotropic curve and c_2 is its conjugate complex curve (where the three imaginary coordinates are reversed), then you will get a translation surface in six-dimensional space (which obviously surpasses our imagination). If you ignore the imaginary coordinates among the points of this surface, then the real portion in three-dimensional space will become visible as a minimal surface!

There is an associate curve for each isotropic curve

Let there be an isotropic curve m with a corresponding minimal surface M. Then you can show that another "associate" isotropic curve can be gained by multiplying the coordinates of the points on m with the complex number e^{it} (i is the imaginary unit and t is a real number). For this new curve, there is once again a minimal surface M_t that is associated with M. And now the big reveal: All these infinitely many associate surfaces have the same surface metrics as M and can, therefore, be twisted into each other without stretching or compressing the surface. The image series on the left shows the so-called Enneper surface being twisted into its associate surfaces.

This is not only fascinating from a "purely geometrical" point of view – there is hardly a lecture on differential geometry that does not discuss, for instance, the example of a catenoid being twisted into a helicoid (top image serids on the right-hand page) – but on p. 134 it will also offer interesting insights in the field of biology.

Demo videos
http://tethys.uni-ak.ac.at/cross-science/enneper-bending.mp4
http://tethys.uni-ak.ac.at/cross-science/catenoid-bending.mp4
http://tethys.uni-ak.ac.at/cross-science/spiral-bending.mp4

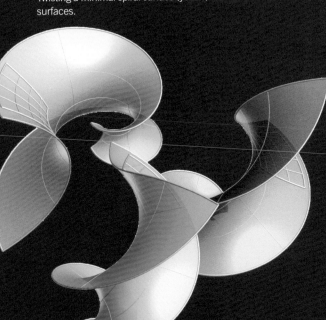

Twisting with unchanged metrics: A two-dimensional organism "living on this surface and missing a flowerbed" will not notice any changes to the surface: for the "flatlander" the flowerbed will always remain the same!

Above: Twisting a catenoid (yellow) into a helicoid (orange). Below: Twisting a minimal spiral surface (yellow) into various other associate surfaces.

Creating Minimal Surfaces by Approximation

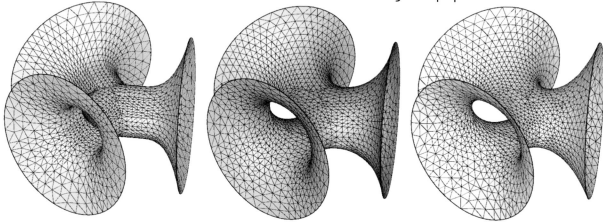

Solution without differential equations

We approximate such a surface roughly through simple surface components (parts of planes, cylinders, or toruses). The components are tessellated (triangulated or otherwise divided into polygons) in such a manner that the distances between the vertices of the individual facets are approximately equal.

Magnetic points on the approximation net

Every point on the net that is generated in that way is considered to be magnetic. Now we start a converging real-time iteration, which allows the points to move according to the rules of magnetism. Boundary curves or individual points can be fixed and manipulated. The corresponding algorithm is adjusted to previous algorithms by Fruchterman and Reingold.

Approximation of minimal surfaces with boundary curves

The result is an approximation of a minimal surface that is defined by fixed boundary lines. There are three advantages to such surface design:

First of all, it is difficult to find an exact solution to this problem with the help of differential equations. Second, the algorithm works interactively in real time. This means that the designer can change the minimal surface forms almost as quickly as with regular freeform surfaces.

Finally, the surface is already triangulated accordingly, which is beneficial for further processing.

Example above: trinoid – the orange surface is the initial surface (cylinder or torus components). The red and blue surfaces are iteration results with and without surface elasticity. The blue surface is the best approximation of the corresponding minimal surface.

Example below: (variation of the circle radius): This time we already take the approximation generated last (orange) as the starting point.

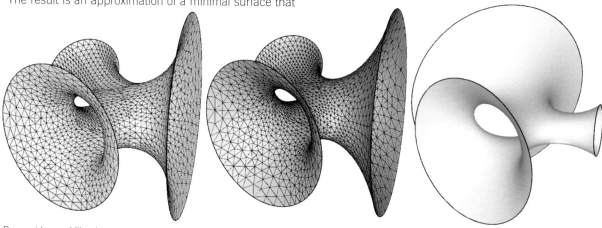

Demo video and literature

http://tethys.uni-ak.ac.at/cross-science/trinoid-variationen.mp4

F. Gruber, G.Glaeser: Magnetism and Minimal Surfaces – a Different Tool for Surface Design

Computational Aesthetics in Graphics, Visualization, and Imaging (2007)

https://diglib.eg.org/bitstream/handle/10.2312/COMPAESTH.COMPAESTH07.081-088/081-088.pdf?sequence=1

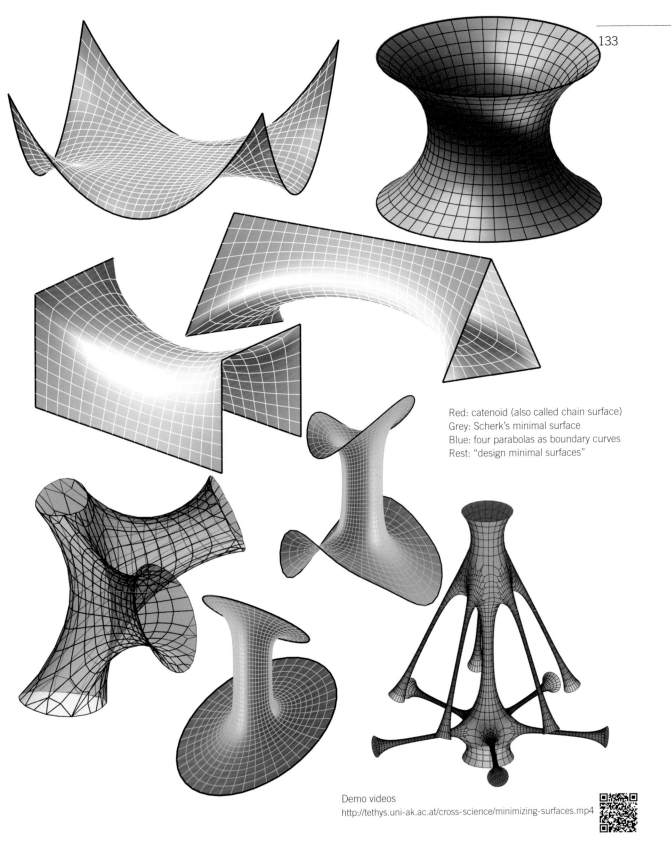

Red: catenoid (also called chain surface)
Grey: Scherk's minimal surface
Blue: four parabolas as boundary curves
Rest: "design minimal surfaces"

Demo videos
http://tethys.uni-ak.ac.at/cross-science/minimizing-surfaces.mp4

Rapid movements in fluids

From a physical perspective, water and air are both fluids. For this reason, very similar physical laws apply to both (consider, for instance, the aerodynamic and hydrodynamic paradox). The denser the fluid and the larger the mobile object, the more visible various effects become to us.

Wings

An insect wing moves, for instance, with a frequency ranging from 50 (for instance, the hummingbird hawk-moth at the top of the right-hand page) to 300 beats (a honey bee). "Ray wings" (below) or the wings of sea snails (image series to the right) might flap up and down only once a second (frequency 1).

Minimising changes on the surface

With squids (sepia, image series to the left), you can clearly see the rapid movements of the wings on the edges of their bodies. How can a living being make such movements without excessive force in a fluid as dense as water? If the distances between the individual points on the wing's surface changed constantly, this would take a lot of effort. However, the wings of the animals pictured here do not seem to stretch or compress at all.

Demo videos
http://tethys.uni-ak.ac.at/cross-science/sepia.mp4
http://tethys.uni-ak.ac.at/cross-science/aplysia.mp4

Ideally minimal surfaces

If the surfaces in question are developable (that is, single curved), then the surface metrics (distances, angles) will be retained automatically. However, if the surface is double-curved, then there will always be folds and wrinkles even when bending the surface slightly.

Only if the surface comes close to a minimal surface, then its isometric bending will work effortlessly without any impractical hydro- or aerodynamic distortions.

And, lo and behold: nature always gravitates towards this solution. This can be seen very nicely with brown algae as they dance in the waves: their surfaces are very close to minimal surfaces (images to the right, which also include a computer-generated minimal surface for comparison).

Demo videos
http://tethys.uni-ak.ac.at/cross-science/taubenschwaenzchen.mp4
http://tethys.uni-ak.ac.at/cross-science/braunalgen.mp4

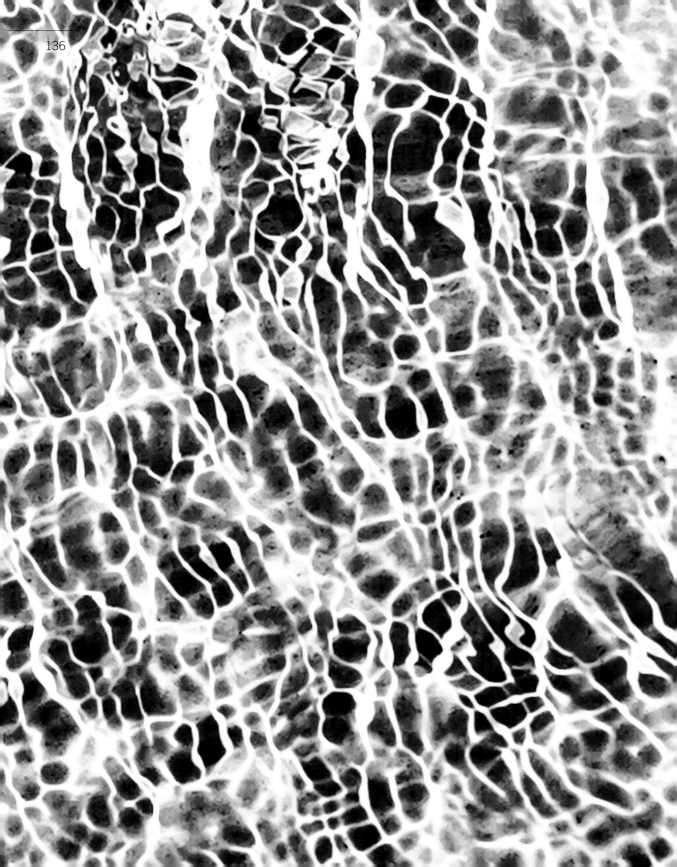

Wave Models:
Strange Phenomena

Reflections of a Water Wave

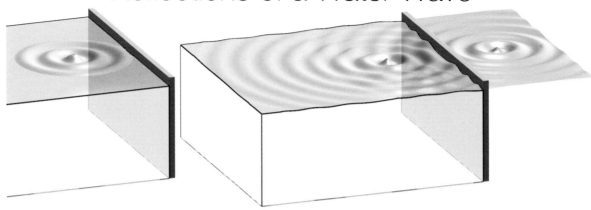

A wave in a swimming pool

A point-shaped excitation source generates a wave that spreads in a circular manner (top left). Once the wave hits a wall, it makes sense from a geometrical point of view to take the virtual reflected wave into consideration in order to calculate interferences (top right). When the wave hits a second wall that is vertical to the first, then things become more complicated. It is clear that the second wall generates an additional virtual counter-wave that disrupts the wave pattern in the pool. Moreover, the parts of the wave that reach a corner are reflected twice (like a ball on a pool table). The impact of this phenomenon can be determined by considering a third virtual wave that is a secondary reflection of the reflected waves on both walls (image below).

Although water waves do not have the same characteristics as light waves, we can approach the problem of multiple wave reflections in a similar manner as we will later when dealing with multiple miror reflections on p. 226.

Demo videos
http://tethys.uni-ak.ac.at/cross-science/wave-reflections.mp4
http://tethys.uni-ak.ac.at/cross-science/wave-interferences.mp4
http://tethys.uni-ak.ac.at/cross-science/waves-in-pool.mp4

Wave patterns at the bottom of the pool

In the animation of overlapping waves, we can, for a given light source, add the patterns in real-time that are projected onto the bottom of the pool. The underlying idea is the following: for an amount of points on the pool bottom, we calculate the corresponding point on the wave surface including the surface normal. On this normal, the corresponding light ray is refracted and intersected with the base plane. The ground surface is illuminated in the vicinity of this point. Quite remarkably, this process will generate net-like structures like the ones pictured here in the large image. The first video of the sea's sand bottom shows how fast this pattern changes in reality. The second video shows that when sunlight hits the water at a flat angle, we get these bright spots due to the different refractive indices of the individual spectral colours.

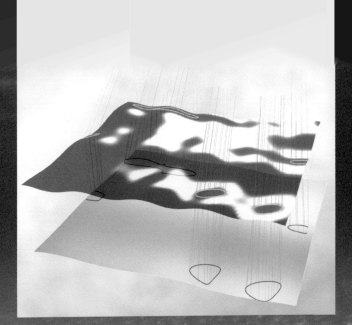

Demo videos
http://tethys.uni-ak.ac.at/cross-science/light-patterns-in-water.mp4
http://tethys.uni-ak.ac.at/cross-science/rainbow-colors-under-water.mp4

Light Diffraction on Double Slits

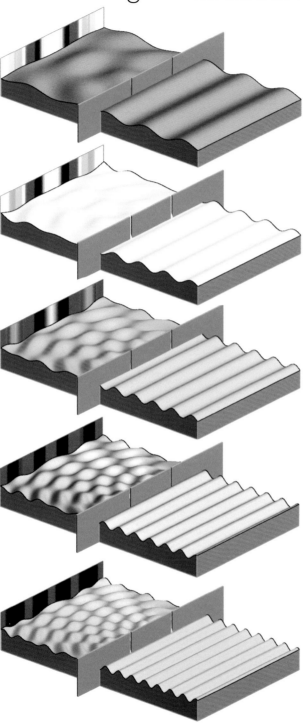

The wave theory of light

Now we know: Light behaves as both waves and particles at the same time. We can find proof for its wave characters in the double-slit experiment: If you direct, for instance, a laser ray (light with uniform wave length) through two narrow parallel slits – the distance between the slits should not be shorter than the wave length of the light – onto an observation screen at an appropriate distance, then you will get interference patterns that are shaped like "bar codes". These are due to interference effects, which are, in turn, the result of light diffraction on the two slits.

Depending on the wave length…

…the shape of the bar codes will differ and there can be overlaps. The image series on this page shows how with decreasing wave length – from red to yellow, green, blue, and violet – the diffraction will generate increasingly detailed wave patterns.

Since sun light, for instance, consists of an entire spectrum of such wave lengths, the resulting pattern on the screen gets increasingly complex. In this concrete case, the distance between the slits was only slightly wider than the wave length of red light (700 nanometres, that is, slightly less than one thousandth of a millimetre).

Varying parameters

In images 1 and 3 on the right-hand page, the distance of the double slit has been increased for different wave lengths (unproblematic). In images 2 and 4, the distance has been shortened. With long wave lengths (red), the requirements for interference are thus no longer met and no pattern will appear on the screen. There is also no pattern when only one slit is materialised (images 5 and 6). The final two images show how – with a valid distance between the slits – the trough (minimum point, image 7) and crest (maximum point, image 8) of the wave are manifested on the observation screen.

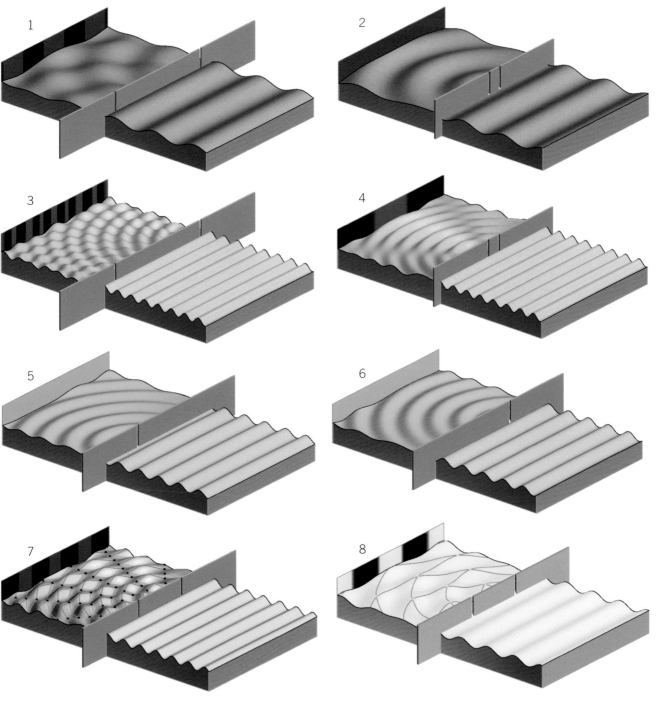

Demo videos
http://tethys.uni-ak.ac.at/cross-science/doppelspalt1.mp4
http://tethys.uni-ak.ac.at/cross-science/doppelspalt2.mp4

A soap bubble reflects its surroundings. However, we also see a lot of "false colours". Getting to the bottom of this will eventually enable us to interpret the colours of peacock feathers (next double page) in new ways.

Thin layers that reflet twice

Let us consider the image at the top right: A locust is sitting on a thin glass plate. We see not only one but two mirror images. How does this happen? If light travelling through air hits a separating layer bordering on a medium with a higher refraction index (here: the upper side of the glass plate), it will generally lead to two things: Part of the light is reflected on the separating layer, and the other part is refracted towards the normal. If the medium is only contained within a thin layer (here: the glass plate might only be two millimetres thin), then there will be another reflection and refraction on the other side (this time away from the normal, centre right image). The light reflected on the bottom side now travels back to the upper side of the separating layer, where it is again (after being refracted away from the normal) partially refracted back, exiting parallel to the light that has been reflected on the upper side. This is why we see two "parallel" mirror images.

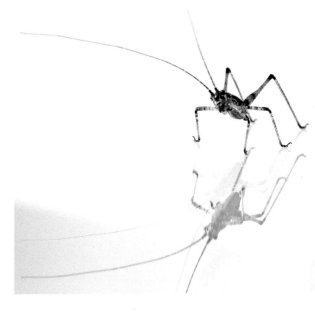

Extremely thin layers

The walls of soap bubbles are *extremely* thin layers. The effect described above can also be encountered there, but the two mirror images are only marginally different (which is why the reflection on the soap bubble of the left-hand page appears in sharp focus). If the wall of a soap skin is less than a micrometre thin (that is, corresponding approximately to the wave length of light), then the distance between the two mirror images is small enough to produce interferences between the two parallel light rays (we will examine this in more detail on the next double page).

The wall thickness of soap bubbles varies continuously due to gravity (until the bubble bursts at the thinnest spot). As a result, the interferences keep varying. Soap bubbles that are too thin or too thick will have no colours.

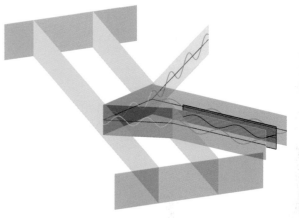

Sometimes there will be additional effects

The photograph to the right was taken from an image series where the effect was not clearly visible: In the intersection area of two soap bubbles, there is a reinforcement of the shimmering effect that is likely produced because the shimmering bubble in the background must first send the double mirror images through the bubble in front of it. This superimposition of thin layers will also be discussed on the next double page.

Demo video
http://tethys.uni-ak.ac.at/cross-science/bubbles.mp4

Magnificent Colours without Pigments

From nature's bag of tricks

Some surfaces in the animal kingdom appear to shimmer in many colours. Yet, it can be shown: not all of these colours are derived from colour pigments. Sometimes you get "iridescent colours", similar to the colours on soap bubbles – but they are often even more "artful".

The jewel beetle (pictured to the left), which is smaller than a centimetre, clearly deserves its name with its bright green, gold, and red colours. However, if you find this beetle in the wild, on a flower or a leaf, it will appear unremarkably single-coloured, actually almost black. It is only the camera's flashlight that will conjure the beetle's bright colours. A similar phenomenon can be found in the remarkable patterns on the eyes of horseflies (bottom right). The different horsefly species will recognise each other by their distinct patterns. Some of these patterns are actually invisible to us, because they fall within the ultraviolet spectrum.

The colours of peacock feathers

The colours of peacock feathers (bottom left), on the other hand, can be seen quite clearly from most viewing angles. We will get to the bottom of this on the right-hand page.

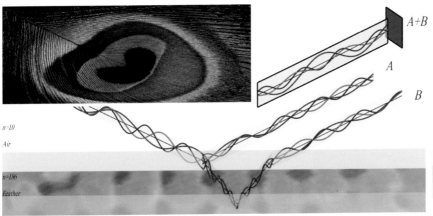

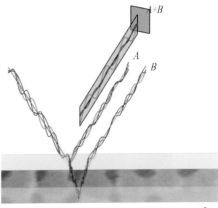

$n=1.0$

Air

$n=1.86$

Feather

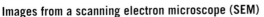

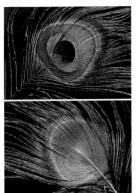

Extremely thin superimposed layers

Top left: we move a computer mouse over an arbitrary point on the displayed peacock feather and will see how each colour displayed is produced with several superimposed nanolayers of varying thickness through refractions, reflections, varying angles of incidence, and resulting intereferences when the light exits (for variations of this, see the image series to the right).

The images on the left show one and the same peacock feather, once from the front and once from behind.

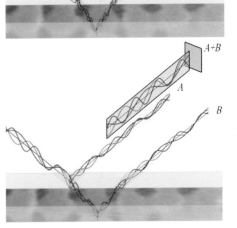

Images from a scanning electron microscope (SEM)

The images below are SEM micrographs of the scales on the wing of a famous blue morpho. An extreme magnification (right) reveals the nano layers (Science Visualization, University of Applied Arts Vienna: Rudolf Erlach, Alfred Vendl).

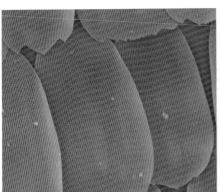

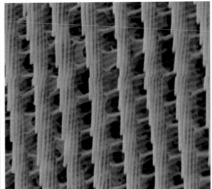

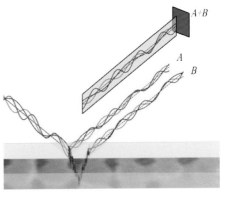

Demo video

http://tethys.uni-ak.ac.at/cross-science/peacock.mp4

Reflecting CD

Rapidly changing colour patterns

Most of us will have wondered about the strange reflections on a CD. Two things stand out here: first of all, the reflections are glowing curves that change their positions rapidly (basically rotating around the centre of the CD), and second, you can clearly see all rainbow colours. The CD tracks (it is actually just one track) have nothing to do with this: the effect can also be observed on blank CDs.

Polycarbonate and aluminium layer

Several characteristics of light are at play here – aside from the mere reflection. Loosely speaking, we are dealing with an interference effect here. A CD is essentially made of a plastic called polycarbonate, which is vaporised on one side with a layer of aluminium (the aluminium reflects the laser beam which reads the data). To avoid chemical reactions, the disc is coated with transparent plastic.

Complicated beam path

When a light beam tries to pass through the plastic coat, part of the beam is reflected on the coat's surface. The rest is refracted on the surface depending on the angle of incidence (and thus separated into the different colour components). On the opposite side of the coat, or, at the latest, on the aluminium layer, the light is reflected, that is, "refracted back", and the beam comes out parallel to the beam reflected on the first surface, though slightly offset.

Optical difraction grating

On top of that, the light beams are difracted because there is an optical difraction grating (several thousand openings per centimetre). The reflected light beam and the light beam that, after some wandering around, is slightly offset in a parallel direction are now interfering. This can lead to the intensification of whole colour components, but also to their erasure. We might already be familiar with this effect from peacock feathers. The image on the right-hand page was shot in a room on the top floor. The sun shines through a window on a slanted wall and partially illuminates one half of a CD that has been positioned at an inclined angle. One might expect only the illuminated half of the CD to create a reflection on the slanted wall. In reality, the whole CD is reflected on the wall, but the reflection is perfectly divided into two halves that also look differently.

Especially the image on the right shows how theoretical-geometrical considerations are "overwritten" here by complicated physical relations.

Demo video
http://tethys.uni-ak.ac.at/cross-science/cd-color-play.mp4

Demo video
http://tethys.uni-ak.ac.at/cross-science/cd-reflection-on-wall.mp4

Koboldmaki *(Tarsius spec.)*

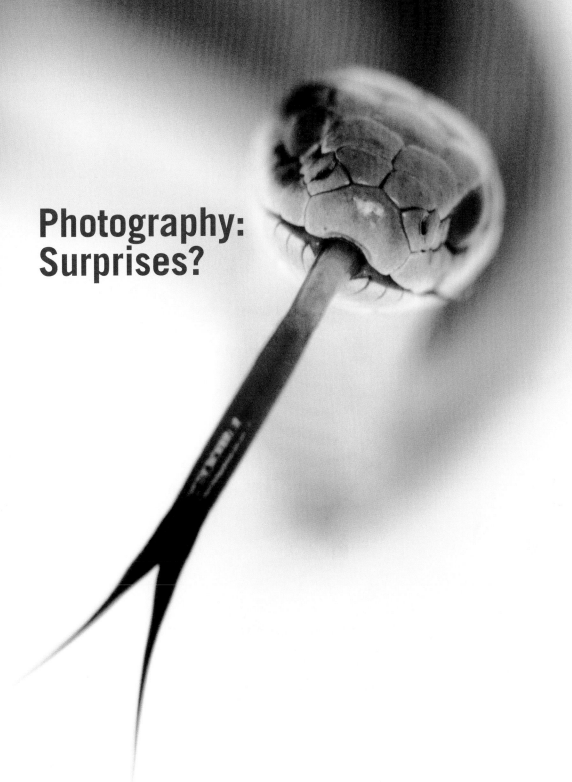

Photography:
Surprises?

Lens Systems

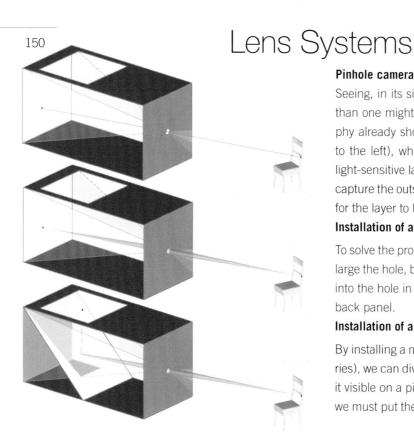

Pinhole camera

Seeing, in its simplest form, is actually less complicated than one might think. The first experiments in photography already showed that a light-proof box (image series to the left), which had a small hole on one side and a light-sensitive layer on the opposite surface, is enough to capture the outside world. However, it must be very bright for the layer to be sufficiently exposed.

Installation of a convex converging lens

To solve the problem of insufficient exposure, you can enlarge the hole, but then you must install a converging lens into the hole in order to concentrate the light rays on the back panel.

Installation of a mirror

By installing a mirror into the box (lower images of the series), we can diverge the generated image and thus make it visible on a piece of wax paper on the box lid. (For this we must put the box and our head beneath a dark cloth.)

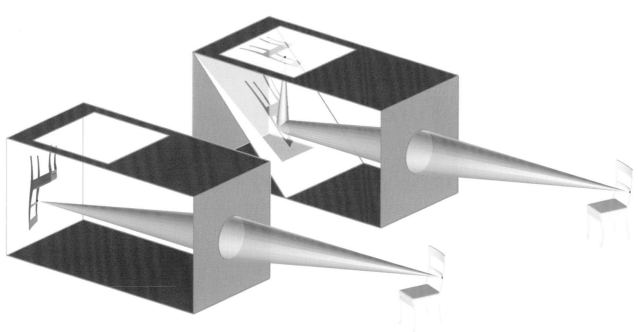

Demo videos
http://tethys.uni-ak.ac.at/cross-science/pinhole-camera.mp4

To what extend is one-eyed vision simulated here?

For a long time, people were convinced that perspectival images produced by a pinhole camera (and generally by any common camera by any manufacturer) were practically identical with the images that we – with one eye – perceive of the three-dimensional world. This might be true to a certain extent, but we should still take a closer look at this.

Lens systems instead of simple convex lenses

The top right image shows the complex light ray trajectory through a modern telephoto lens: the – almost parallel – light rays marked in green, which emanate from a distant point P, are refracted 20 times and more until they converge on the chip into image point P^c. All these light ray trajectories could theoretically be replaced by a single ray through the lens centre Z (marked in red in the image).

Aperture

In photography we also have the problem that a sufficient amount of light must reach the sensor plane. This is achieved through the nearly circular aperture, which can be opened to varying degrees (second image to the right). As we will see on the next double page, opening the aper-

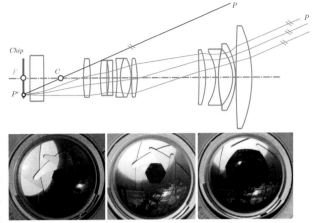

ture has advantages and disadvantages: a large aperture (small f-number) allows the photographer to capture astonishing details at the expense of depth of field, which is often preferred for more artistic shots. Small apertures, on the other hand, enable – up to a certain physical limit – greater depth of field.

Lens eyes are also much more than a pinhole camera

The images below illustrate how, after extreme refraction on the retina, the lens of the eye is in charge of "fine tuning".

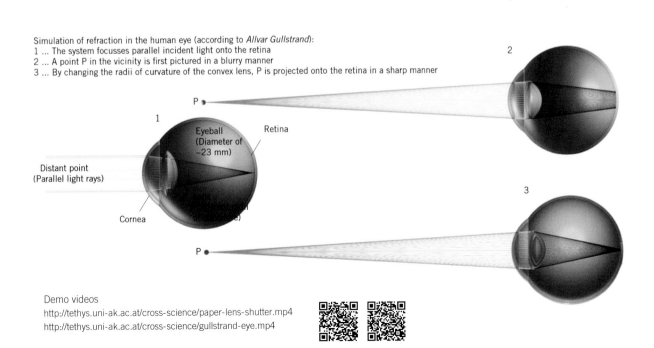

Simulation of refraction in the human eye (according to *Allvar Gullstrand*):
1 ... The system focusses parallel incident light onto the retina
2 ... A point P in the vicinity is first pictured in a blurry manner
3 ... By changing the radii of curvature of the convex lens, P is projected onto the retina in a sharp manner

P

Eyeball
(Diameter of
~23 mm)

Retina

Distant point
(Parallel light rays)

Cornea

P

Demo videos
http://tethys.uni-ak.ac.at/cross-science/paper-lens-shutter.mp4
http://tethys.uni-ak.ac.at/cross-science/gullstrand-eye.mp4

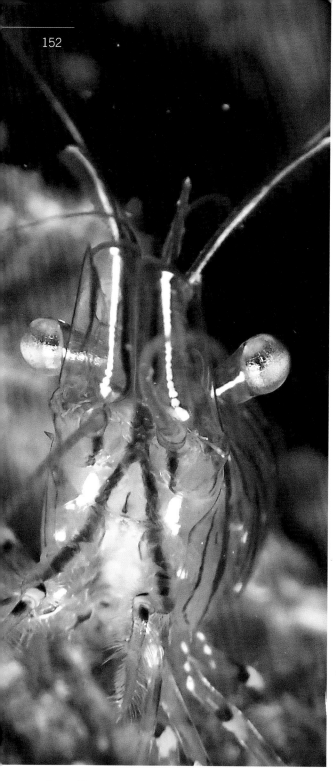

Three Eye Types

Lens eyes and compound eyes

These two eye types are widely spread in the animal kingdom. Especially lens eyes are well familiar to us (p. 148). Invertebrates often have compound eyes (right-hand page at the top, though there is a stalk-eyed fly on the right), but there are also some with lens eyes (right-hand page at the bottom). The close-range vision of both eye types is remarkable, which is relevant to these animals. With compound eyes, several facets will work together like a single lens.

"Lobster eyes"

From a geometrical point of view, there is a third remarkable eye type, which has established itself with lobsters and shrimps: with these eyes the incident light rays are collimated, not through a lens, but through multiple reflections on mirrored prismatic facets. The first video shows multiple reflections in a cube corner; the second shows how a light ray moves through a single square prism; and the third shows how prism arrangements collimate many individual light rays on the retina.

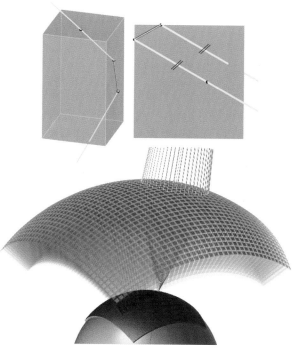

Demo videos
http://tethys.uni-ak.ac.at/cross-science/reflecting-corner.mp4
http://tethys.uni-ak.ac.at/cross-science/ray-through-single-prism.mp4
http://tethys.uni-ak.ac.at/cross-science/rays-through-many-prisms.mp4

Elephant and Fly Photography

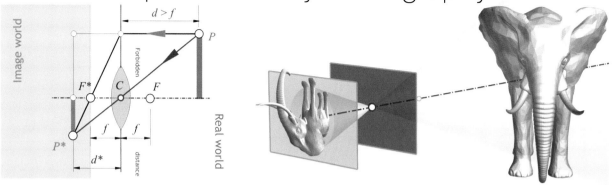

Photography does not produce even images

In texts about the photographic image, we often find the following simplification:

points in space are projected from the lens centre onto a plane (the sensor plane). However, if you look at the sketch at the top left, which takes into accounts the physical laws of lenses, you will immediately see that only points with the same distance from the "main plane" through the lens centre "land" on one plane. If you apply this sketch to all points of a spatial object, you will get a virtual three-dimensional object behind the lens.

Photographing large objects

The focal length of a camera lies usually in the centimetre range. If we photograph an object that is several metres away and thus also quite large ("elephant photography"), then the corresponding virtual object will end up being very flat (see top right image). As a result, the object will appear almost entirely in sharp focus.

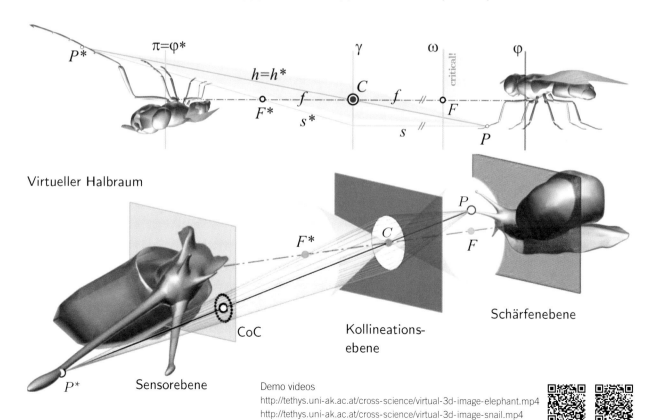

Demo videos
http://tethys.uni-ak.ac.at/cross-science/virtual-3d-image-elephant.mp4
http://tethys.uni-ak.ac.at/cross-science/virtual-3d-image-snail.mp4

When the photographed objects are small

When we photograph small objects like a fly or a snail (images at the bottom of the left-hand page) from a range of only a few millimetres or zentimetres, so that they fill the whole short, then we have to move close to the object. The corresponding virtual object is then no longer flat at all, and as a result, the vast majority of points will appear as blurry, circular discs on the sensor (CoC = circle of confusion). In order to keep this circle of confusion as small as possible, you can use a smaller aperture (stopping down). However, the aperture cannot be too small, because then you might get diffraction effects of the light (diffraction blur).

Big camera, small objects

The images on this page illustrate this problem based on a relatively giant macro camera (focal length of 65 mm) with twin flash, which can produce picture-filling close-ups of objects that measure only a few millimetres. The flash is necessary because otherwise the amount of light passing through the small aperture would not be enough to exposure the sensor.

Approaching the "forbidden plane"

The virtual 3D image will quickly increase in size when we approach the plane that lies before the lens centre at a single focal length. If an object is located at a double focal length from the lens centre, then the expansion of the virtual image in the direciton of the optical axis can already be compared to that of the real object. If the object moves even closer, then the virtual image will quickly become *very* large. Within less than a single focal length, you can no longer capture the object in sharp focus.

The problem of decreasing depth of field is obviously not easy to solve. On the next double page, we want to analyse an at least theoretical solution to the problem.

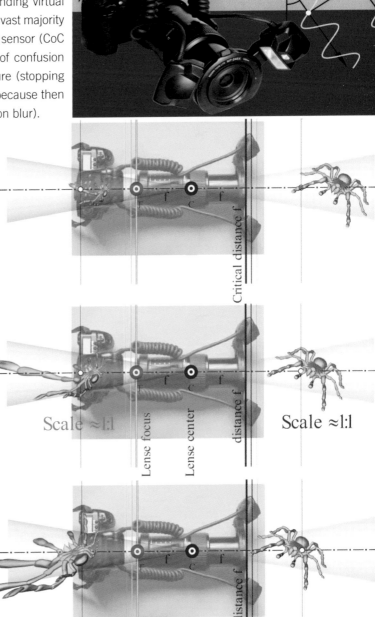

Demo video
http://tethys.uni-ak.ac.at/cross-science/virtual-3d-image-spider.mp4

Sharp Images All the Way Through?

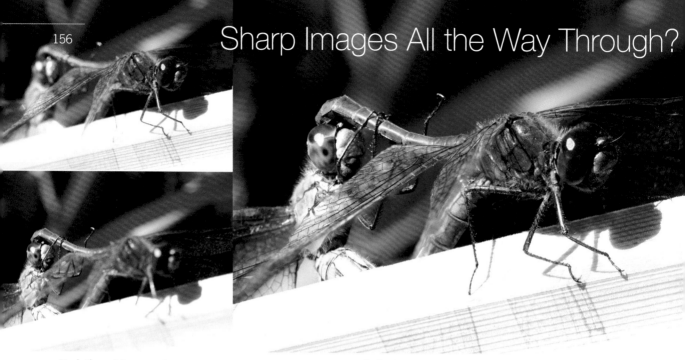

Variation of lens centre

Above: two mating dragonflies which have been photographed several times with varying lens centres (and focal points). The first and the last photograph of the series can be seen on the left. The background remains blurry, which is considered to be beneficial from an aesthetic point of view. "Focus stacking" can thus also be seen as an artistic tool.

Following the conclusions of the previous double page, such photographs cannot be sharp from rear to front. Yet, there is a trick that works at least when insects remain still for a second: now we take several photographs where – with a constant focal length – the distance of the lens centre to the object varies slightly.

The focus plane scans the object

If we superimpose the single frames now and let a software determine where the sharpest points of frames are, then we get a remarkable result (bottom-left image, with the real object visible to the right of the lens centre): the scene appears sharp but distorted.

"Individual scaling" of the frames

Calculations that are derived in the work cited below show that before superimposing the single frames, we must resize the frames to different scales.

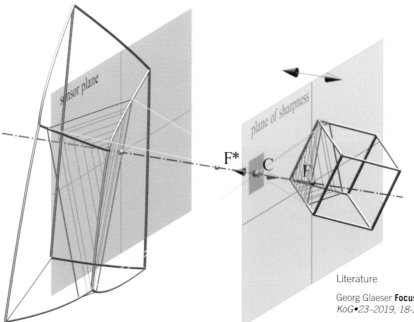

Literature

Georg Glaeser **Focus Stacking from a Purely Geometrical Point of View.** *KoG•23–2019, 18-27 (2019)* https://hrcak.srce.hr/file/335384

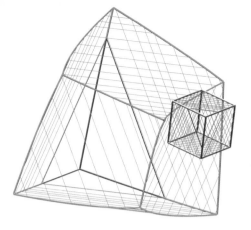

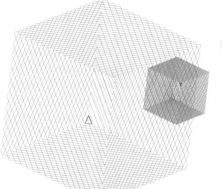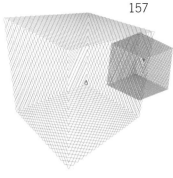

Depending on the scaling factor...

We can apply different sets of scaling factors. For one, we will get – as expected – perfectly sharp perspectives (top-right image), but we can also apply scaling factors that will yield a completely sharp orthographic projection. This is remarkable because such projections are actually impossible to achieve with a single frame: even with extremely powerful telephoto lenses, we get perspectival distortions. The benefit of orthographic projections is enormous from a purely geometrical point of view: parallel edges remain parallel and are all shortened to the same extent. This enables the accurate measuring of small objects.

Artistic aspects

On the right, eight frames have been stacked (flower of a dog rose). On the bottom right, only three frames have been stacked, so only significant parts of the praying mantis appear in focus. The blurry background of the stacked photograph is intentional.

Again: the object of desire must remain still at least long enough for the single frames to be saved.

Demo videos
http://tethys.uni-ak.ac.at/cross-science/focus-stacking1.mp4
http://tethys.uni-ak.ac.at/cross-science/focus-stacking2.mp4
http://tethys.uni-ak.ac.at/cross-science/focus-stacking3.mp4

Capturing Space on a Sphere

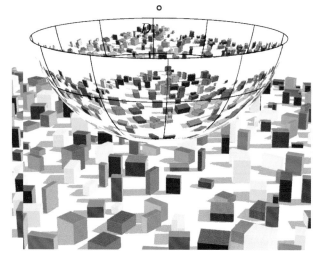

Systematically recording an entire space

Imagine you are launching a drone to point D (top left image) and remain in this position. By panning and tilting, you take a sufficient amount of photographs from this position (top right image) – modern drone models are very well capable of doing so. Now imagine a sphere with an arbitrary radius around the point D (middle left image) and project all image points from the entirety of image points onto this sphere. Many points will appear multiple times, but obviously each point must be recorded only once (as a pixel on the sphere).

Looking through a virtual reality headset

Now, at some later point in time, look at the result on a computer through a virtual reality headset. You can (without changing your position) twist and turn your head in any direction or even rotate in a full circle. In doing so, you will get *exactly* the same impression as an observer on position D during the moment of recording.

Faking reality without restrictions

Optically, you can, in fact, not distinguish whether you are actually on position D or not. If the hardware of your computer is specialised on such projections, they can be done in absolute real time (bottom left image).

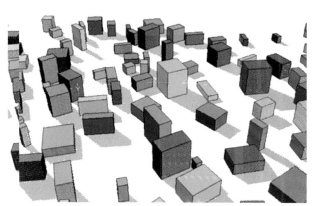

Demo video
http://tethys.uni-ak.ac.at/cross-science/spherical-drone-view.mp4

Practical test

To verify this theory, we will implement the process described on the left-hand page in practice: a drone has been manoeuvred to a fixed point, and about 30 shots of the environments have been taken and transformed into pixels on a sphere (top images). Since the images on the sphere overlap, they have been cropped along great circles and small circles of the sphere (left image). Now, with a virtual reality headset (or, using a suitable software, with a mouse or cursor buttons), you can once again "put yourself into the position of the drone" and view the world from there, without optically noticing any differences to reality.

Why does this work?

With only one eye, we cannot measure any distances. Our optical system is based on angular measurements, which matches in both cases. It is remarkable that we even perceive straight lines as straight lines. On the sphere, these straight lines first appears as segments of great circles, which, in turn, appear as straight lines when viewed from the sphere centre (consider the image at the bottom of the left-hand page).

Demo videos
http://tethys.uni-ak.ac.at/cross-science/spherical-drone-view-reality.mp4
http://tethys.uni-ak.ac.at/cross-science/spherical-drone-view-reality2.mp4

Strange Photographic Results

An annoying effect in high-speed photography

The so-called rolling-shutter effect is an effect that is well-known in the industry and that can occur with photographs and video recordings of fast moving objects. Most cameras cannot expose the entire sensor to the light all at once; instead you should imagine a "line of simultaneous exposure" that wanders over the object either row by row or column by column, which takes a short but not insignificant amount of time.

Can you rely on a photograph?

The almost artistic-looking photograph on this page shows what can happen when, for instance, a red longhorn beetle (*Corymbia rubra*) launching into flight is photographed despite an extremely short exposure time (1/2500 s): the body of beetle "barely" moves, but the elytra move no-ticeably faster than the body (they vibrate with the wings during flight!). The wings themselves oscillate at a very high velocity in relation to the body.

What is the cause of this effect?

The shutter speed of a camera is for most camera models considerably longer than the extremely short time that is typically specified. Since the exposed sensor is recorded line by line and the recording of each line takes only a fraction of the shutter speed, the exposure time of a processed line is, therefore, much shorter. However, until the electronic elements reach the next line, a (very short) time span has passed, and in the next line – once again exposed for an extremely short period – a new situation is recorded: in the meantime the beetle's wings, for instance, will have already travelled a certain distance. This is why we get these stranges bends, loops, and wipes.

A drone is "standing" above the observer and is photographed with a shutter speed of "1/16 000"-second twenty times per second. The camera was held in portrait orientation.

Demo videos
http://tethys.uni-ak.ac.at/cross-science/drone-hd.mp4
http://tethys.uni-ak.ac.at/cross-science/drone-hispeed.mp4

The Problem with Closeness to Reality

The following rule of thumb applies during flight: The wings must always move symmetrically. If this is noticeably not the case in a photograph, you can assume the cause to be the rolling shutter effect. There are, however, dead-point positions during flight which allow you to take realistic pictures of flying animals.

Here some wing-beat frequencies for those who are interested:
Birds: from 1 Hz to 78 Hz (griffon vulture, amethyst woodstar).
Insects: from 20 Hz to 300 Hz (desert locust *Schistocerca gregaria*, bees, flies).
Smallest fungus gnats (Mycetophilidae): possibly over 1000 Hz.
Bats: from 7 Hz to 18 Hz (megabats, lesser mouse-eared bat).

Image error or not?

Image errors are not always as obvious as on the previous page, where a drone is pictured. They can look so true to life that one might be tempted to take them at face value. However, especially in high-speed photography, you should take supposedly realistic images with a grain of salt! You can apply certain tests to verify whether there are any image errors or not.In the top left image series (6 photographs), a sparrow has been filmed during landing at a relatively low resolution with a high-speed camera (global shutter). For comparison, the top right image series (4 photographs) shows a sparrow launching into flight, which was photographed at a high resolution. Due to the bird's slanted position in the two images at the top (right-hand page), the rolling shutter effect is clearly visible. The two images at the bottom (top right) show dead-point positions of the wings and barely have any distortions – the wings appear to be symmetrical.

Due to the undistroted symmetry, the bird's launch into flight in the bottom image series seems to have been captured correctly as well.

Ways out of this dilemma?

To reduce distortion effects, we can significantly reduce the image resolution. If you film, for instance, only in HD (1920×1080) instead of 4k (Auflösung 3840×2160), each image can be saved four times faster, which greatly reduces distortions – though the details of the images will no longer appear in razor-sharp focus (p. 161 at the bottom).

The other option would be to dig deep into your pocket and splash out on a camera that has "global shutter": with these cameras, all pixels are read out simultaneously. This process is technically very complex, which is why such cameras are quite expensive.

Why does it (usually) work with flash?

An electronic flash illuminates a scene for an extremely brief period, namely for a fraction of a millisecond (e. g. 20 microseconds). When there is little ambient light, the following trick works: you use a small aperture setting and expose e. g. for a relatively long duration of $1/200$-second. This is good time span to fire off a targeted flash, irrespective of whether the flash occurs at the beginning or at the end of this time span or somewhere in between.

Now the sensor is exposed to light for this whole time span, without a slit moving from top to bottom. If the surroundings are dark, then the sensor is barely exposed to light "most of the time" – except for the extremely short period during which the flash illuminates the entire scene. Now you have successfully "frozen" the scene without needing to use a slit.

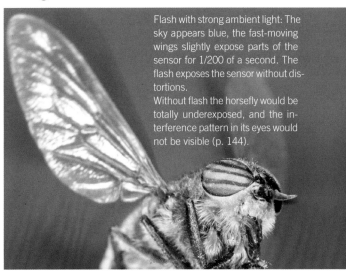

Flash with strong ambient light: The sky appears blue, the fast-moving wings slightly expose parts of the sensor for 1/200 of a second. The flash exposes the sensor without distortions.
Without flash the horsefly would be totally underexposed, and the interference pattern in its eyes would not be visible (p. 144).

Simulation with Computers

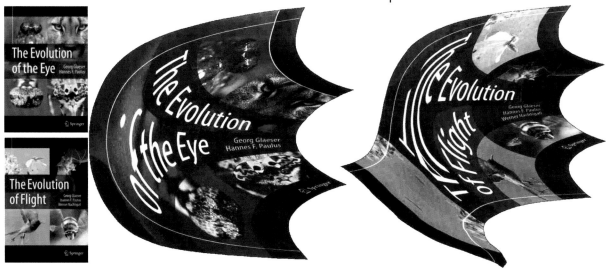

A suitable comparison with scanners

When a photocopier or scanner saves an image, a rod-like sensor inside the machine moves steadily from left to right. In every position the sensor scans a very small rectangular part of the image. All scanned rows are eventually merged into the final image. The scanned image is usually static, because the scanner works like a camera.

Movement of the scanned object

Since the scanning process takes a while, one might come up with the idea of moving the original image while it is being scanned. In the top centre and top right images, the covers of two books have been quickly rotated during

scanning. The result is quite remarkable and very different from the original. In the top right image, the cover has been rotated so quickly that some parts appear multiple times (e.g. look closely at the letter "T" in the title).

Simulation with a computer

The image series below shows a corresponding computer simulation with striking similarities: A rectangle is rotated with varying angular velocities while a scanning line is moved from top to bottom. All points of the rectangle that overlap with the scanning line appear in the final image. The rotational velocity will determine the image's distortion.

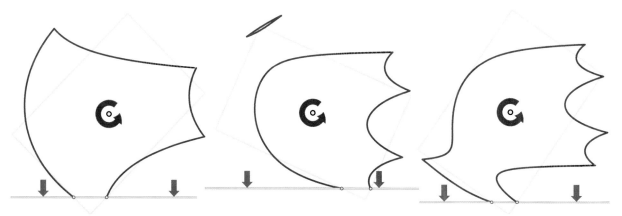

Demo video
http://tethys.uni-ak.ac.at/cross-science/rolling-shutter-rect.mp4

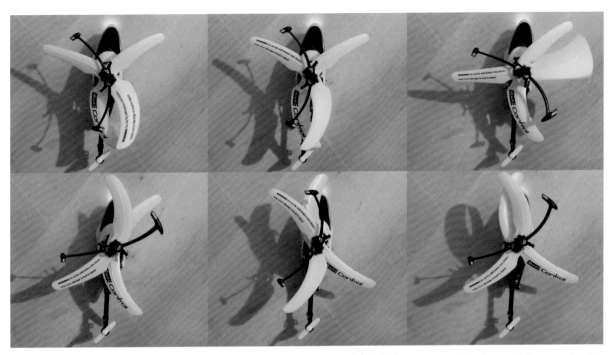

Simulation of a helicopter's rotor blades

To provide another example of a real application of this phenomenon, we have taken several photographs of a miniature helicopter that is the size of a giant dragonfly and has very fast-moving rotor blades. Although the photographs were taking with an extremely short exposure time of $1/32000$ s, the rotor blades, which are actually symmetrical and not bent, appear distorted! (The labels on the rotor blades, on the other hand, are clearly legible.) The computer simulation below illustrates the same process and result.

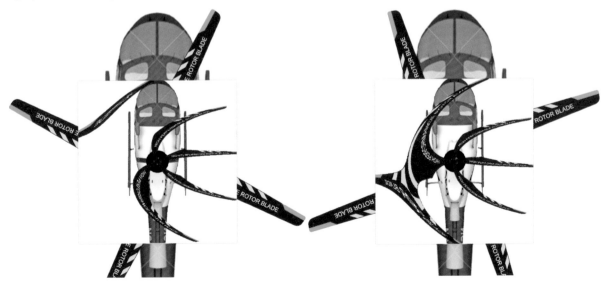

Photo Finish – When There Is a Lot at Stake…

Something is odd about this photograph…

First let us look at a famous photo-finish image taken in the year 1953 (the photographer is unknown): it records the finish-line crossing of three sulkies in a harness race. It is obviously an analogue photograph. Yet, it is still an unusual photograph, because the circular wheel of the sulky at the bottom is represented "incorrectly" – the bottom sulky is also noticeably shorter (we will later see that it must have been the fastest sulky while crossing the finish line).

A photo-finish camera is no conventional camera

In a typical photo-finish camera, the film strip is moved at a certain speed behind a fixed, centred vertical slit. The limited field of vision through the slit aligns with the finish line, so that the line is (photochemically) exposed to the light *at different times*. A photo-finish image is thus not a momentary record, but it allows us to determine at what point in time a certain point crossed the finish line by measuring that point's distance from the finish line. In the past, it took quite a long time to get to that final decision because the negative had to be developed first.

Finish-line crossings in running competitions

The image at the bottom shows a realistic illustration of a finish-line crossing during a sprint competition (artist: Markus Roskar). Some of you might say that there is something wrong about this "photo", but it is not a photo in the classical sense: it only captures bodies or body parts once they cross the finish line. After that they are – in proportion to the time that has passed – displaced to the left. Now look at the calves of the two runners at the bottom. Compared to the rest of their bodies, they seem to cross the finish line a bit later. Since the heels were photographed exactly on the finish line, we can conclude that the runners landed with these feet exactly on the finish line and their heels and calves thus "remained stuck" there for longer.

Interesting video

https://www.eurosport.de/leichtathletik/em/2022/leichtathletik-em-irres-foto-finish-uber-110-m-hurden-eine-tausendstel-entscheidet_vid1731092/video.shtml

Image 1

It cannot get more exciting

Image 4

Two racing cars, green (G) and red (R), drive across the finish line. The photo-finish camera only captures the middle line = finish line (in portrait mode). The pixels are recorded from the bottom to the top within a fraction of a millisecond. Image 1 and image 2: The cars have not reached the finish line yet – nothing has been photographed. G seems to be catching up.

The image is recorded from the bottom to the top, and chronologically speaking, R is captured before G. G thus had an extra microsecond to be photographed. This could have given G its one-centimetre advantage (at 100 m/s the car travels 10 cm in one millisecond, but only 0.1 mm in one microsecond).

Image 2

Image 3 already gives us the final decision, even though the photo-finish image is only partly finished. Green seems to be leading by a centimetre. Images 4, 5, and 6, which complete the photo-finish image, now prompt the question: has G really won the race?

Image 5

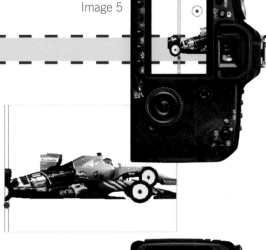

Image 3

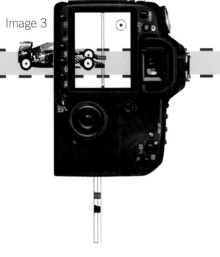

Image 6

Demo video
http://tethys.uni-ak.ac.at/cross-science/photofinish.mp4

Demo videos
http://tethys.uni-ak.ac.at/cross-science/clouds.mp4
http://tethys.uni-ak.ac.at/cross-science/water-turbulences.mp4

Air and Water: Fluids!

Of Wind and Rain Gauges

A sketch by Leonardo da Vinci motivated the simulation of a wind gauge, or anemometer.

Particle method for fluids

The implementation of this idea in programming has proven to be very useful – the software works with the "particle method": air (more generally: a fluid) is simulated with thousands of particles that can meet resistance and thus transmit forces or be reflected (see also p. 172).

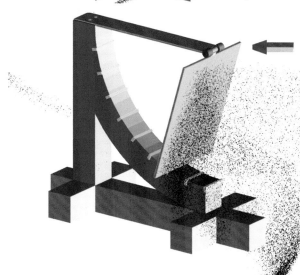

The gauge is not linear

The dial on anemometers cannot be labelled linearly: even with strong wind forces (image below), the direction in which the mobile measuring sheet is blown is only roughly horizontal.

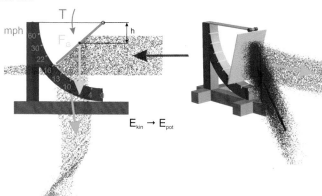

Demo video
http://tethys.uni-ak.ac.at/cross-science/speed-of-wind.mp4

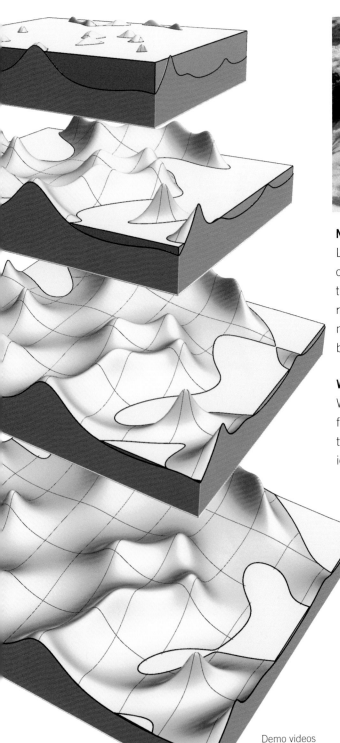

Non-linear dials

Let us look at the rain gauge in the bottom right: it is not cylindrical but tapered off towards the bottom. In addition, it has a cone inside. If we look more closely at the rain gauge's dial, we will see that the measuring lines do not increase evenly. The same can be expected for river beds and even entire landscapes.

When the water level drops

When a lake has an average depth of four metres and its water level drops by two metres, then it will have lost significantly more than half of its volume.

If the foundation of the lake was a cone, then only $1/8$ of the water volume would be left. The image series to the left illustrates the drying process with dropping water levels. It could also be interpreted as an image of a landscape that was covered by a melting glacier (see also the top right photograph).

Demo videos
http://tethys.uni-ak.ac.at/cross-science/evaporate1.mp4
http://tethys.uni-ak.ac.at/cross-science/evaporate2.mp4
http://tethys.uni-ak.ac.at/cross-science/rain-gauge.mp4

Wind Turbines and Water Spirals

Once again: particle method with fluids

On p. 170 we have already seen that the particle method is especially well-suited for simulations for fluidic situations. Each of the thousands of particles is tracked, and the resulting force transmissions are summed up. The complexity of the situation will impact the computing time only marginally.

In this concrete case, we have simulated a kind of wind turbine where multiple parameters can be adjusted: the number of blades, the frictional resistance of the turbine, the flow velocity into the blades, and the position of the jet. It turns out that such simulations are actually quite practible.

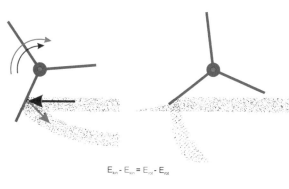

$$E_{kin} - E_{kin} = E_{rot} - E_{rot}$$

Demo videos
http://tethys.uni-ak.ac.at/cross-science/anemometer.mp4
http://tethys.uni-ak.ac.at/cross-science/sprinkler.mp4

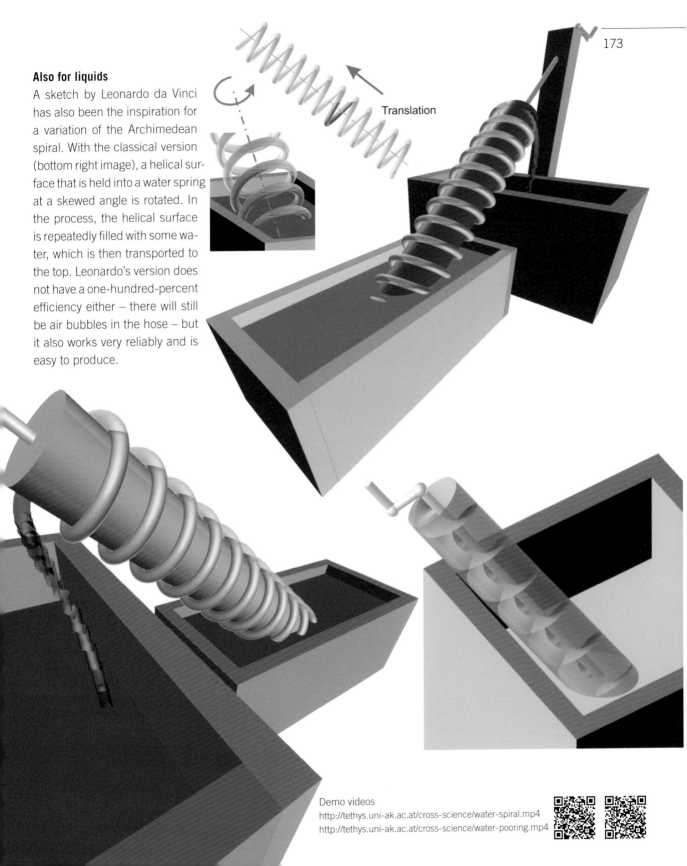

Also for liquids

A sketch by Leonardo da Vinci has also been the inspiration for a variation of the Archimedean spiral. With the classical version (bottom right image), a helical surface that is held into a water spring at a skewed angle is rotated. In the process, the helical surface is repeatedly filled with some water, which is then transported to the top. Leonardo's version does not have a one-hundred-percent efficiency either – there will still be air bubbles in the hose – but it also works very reliably and is easy to produce.

Translation

Demo videos
http://tethys.uni-ak.ac.at/cross-science/water-spiral.mp4
http://tethys.uni-ak.ac.at/cross-science/water-pooring.mp4

Kinetic Sculptures

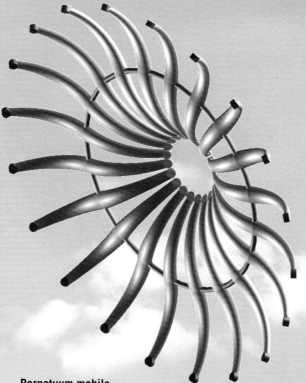

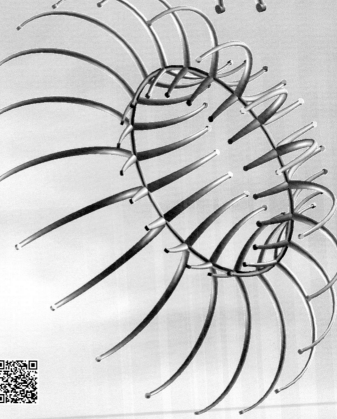

Perpetuum mobile

The American "kinetic sculptor" Anthony Howe creates wind-driven sculptures that can be designed quite well with a computer. One of his best known works is the kinetic sculpture that he created for the Olympic Games 2016 in Rio de Janeiro, Brazil.

Associations

The images on this page will certainly remind you of flowers or – especially if you are into diving – of sea anemones in the water current. The movement on the right-hand page seems to be more complicated, but the motions of the individual ribs are actually just simple rotations around tangents on the central circle. The movement captured in the second video (shot in St. Johann im Pongau / Salzburg) is clearly a more complex spherical spatial motion (superimposition of up to three rotations with intersecting axes).

Demo videos

http://tethys.uni-ak.ac.at/cross-science/anthony-howe.mp4
http://tethys.uni-ak.ac.at/cross-science/kinetic-sculpture.mp4

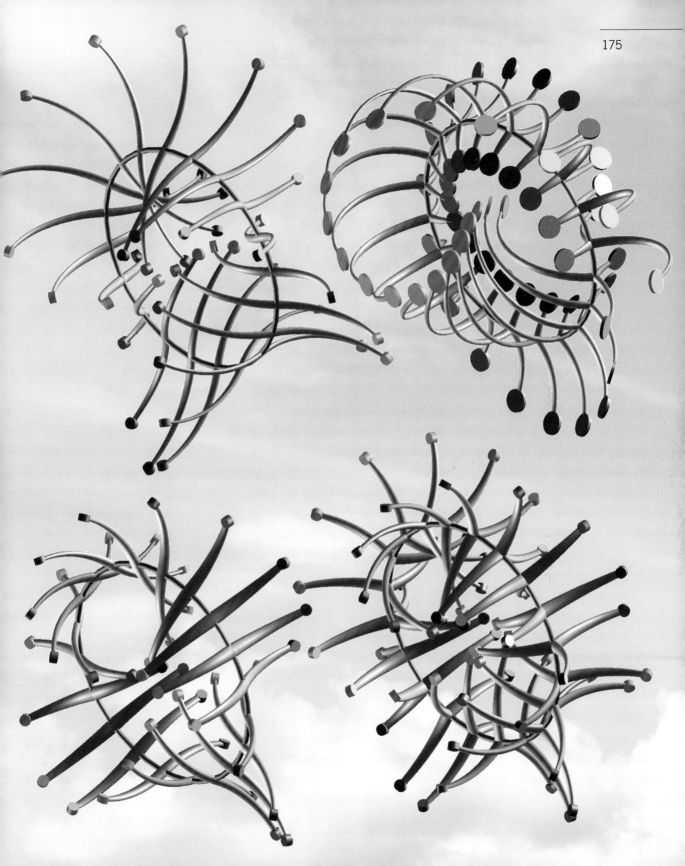

Tube worms...

...use their feathered tentacle crown to sift microplanktom from flowing water. When the worm pictured above, *Serpula*, is disturbed in the process, it will rapidly retract into its living tube made of hard calcium carbonate, closing the tube with a lid – a modified ray of its tentacle crown.

Spirographis swaying its tentacles in the current

The species pictured on the right-hand page, *Spirographis* ("spiral drawers"), is equipped with two tentacle bases, one of which is quite long and carries about 200 to 300 tentacles on five to six spiral windings. The tube, which measures about a centimetre in width, is stuck in mud or attached to a stone; it contains the rest of the worm body that is segmented regularly with 100 to 300 segments.

Flapping lash and bristle crowns with filters

Bristle crowns function like fish traps or paddles. Each tentacle carries many fine appendages with flapping lashes. These lashes swirl water into the tentacle funnel. During the process, very fine particles are filtered, embedded into sticky slime, and transported to the mouth opening. A fish trap is made of bristles that are not too close together. They are meant to let liquid currents with fine particles pass, but not bigger particles. This role is fulfilled by the bristle crowns of the aforementioned worms. They direct the water current in such a way that plankton particles carried by the water are driven to the edges, where they get stuck on very fine setae. The bristle crowns are not swayed back and forth like a fishing net but held still.

Demo videos

http://tethys.uni-ak.ac.at/cross-science/tubeworm.mp4
http://tethys.uni-ak.ac.at/cross-science/spirographis.mp4

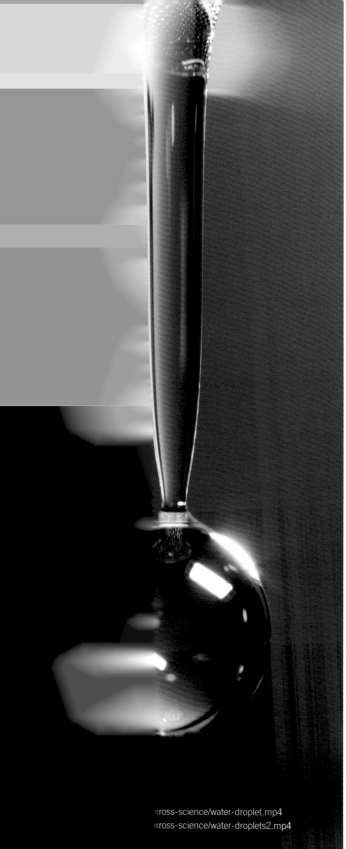

Draining

Cohesion vs. gravity

When a water droplet gathers (for instance, as dew on a spider's net, see top left on the right-hand page), cohesion causes the water molecules to stick together up to a certain size, so much so that small spheres form (cohesion is responsible for the surface tension). On a leaf (right-hand page, second image at the top left), such droplets will, due to the leaf's tilted position, merge into larger – likewise still stable – drops.

If – as with a dripping water faucet – an increasing amount of molecules are added, then cohesion will no longer be strong enough and a drop will tear, but the falling drop – now weightless – will immediately turn into nearly a sphere.

cross-science/water-droplet.mp4
cross-science/water-droplets2.mp4

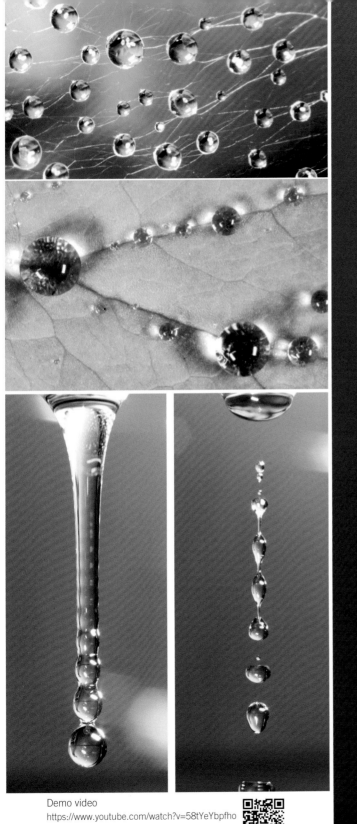

Demo video
https://www.youtube.com/watch?v=58tYeYbpfho

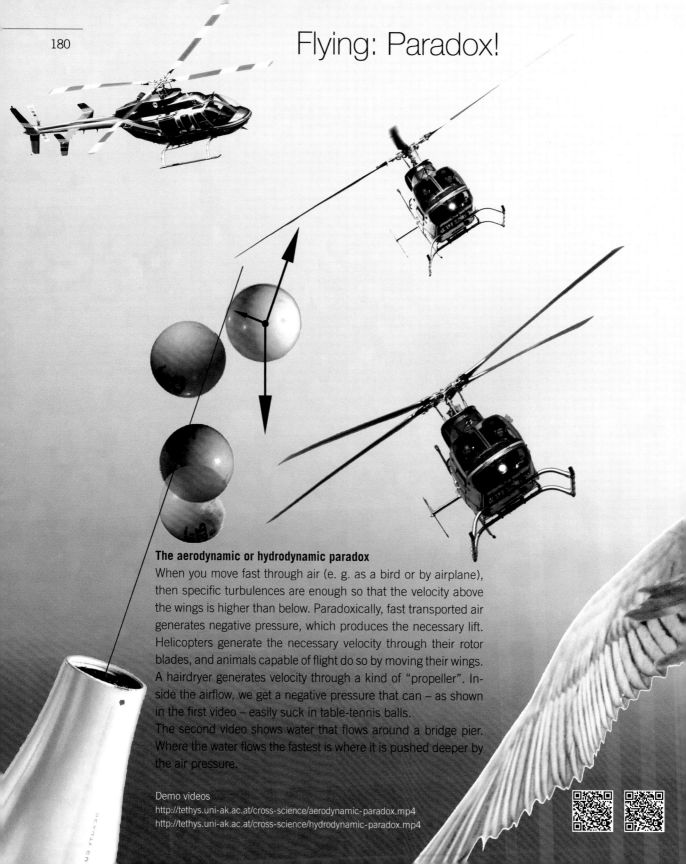

Flying: Paradox!

The aerodynamic or hydrodynamic paradox

When you move fast through air (e. g. as a bird or by airplane), then specific turbulences are enough so that the velocity above the wings is higher than below. Paradoxically, fast transported air generates negative pressure, which produces the necessary lift. Helicopters generate the necessary velocity through their rotor blades, and animals capable of flight do so by moving their wings. A hairdryer generates velocity through a kind of "propeller". Inside the airflow, we get a negative pressure that can – as shown in the first video – easily suck in table-tennis balls.

The second video shows water that flows around a bridge pier. Where the water flows the fastest is where it is pushed deeper by the air pressure.

Demo videos
http://tethys.uni-ak.ac.at/cross-science/aerodynamic-paradox.mp4
http://tethys.uni-ak.ac.at/cross-science/hydrodynamic-paradox.mp4

Stationary and underwater flight

Animals capable of flight, such as birds, bats, and insects, have mastered the way they must twist and bend their wings in order to generate as much lift as possible. Have a look at the two videos cited on this page (the second shows how a penguin "flies" underwater).

Demo videos
http://tethys.uni-ak.ac.at/cross-science/pigeon-shake-flight.mp4
http://tethys.uni-ak.ac.at/cross-science/penguin-flying.mp4

A classical paradox

A paradox is a supposed contradiction that is resolved, however, upon closer inspection. Here we will specifically look at the aerodynamic or hydrodynamic paradox: under certain circumstances, heavy objects can fly or swim.

Before we explore this in more depth on the next double page: An airplane must move very fast while flying; for a big shark that is heavier than water, gentle gliding is enough. A helicopter (or also a drone) must rotate its blades fast, but this allows it to hover motionless in the air or even fly sideways or backwards by modifying the flying object's inflination so that the "lift forces" act on it sideways or backwards.

Demo video
http://tethys.uni-ak.ac.at/cross-science/longimanus.mp4
http://tethys.uni-ak.ac.at/cross-science/vulture-flight2.mp4

Hummingbirds are in a way "insect-like"

They can hover statically in the air without problem and also fly away in all directions. They control the lift generated by the rapidly moving wings (about 50–80 beats per second) by tilting their bodies and through their very flexible tail feathers and steer it in the right direction. A similar strategy can be found with insects, especially with moths from the hawk moth family. In general, the smaller the animal, the more acrobatic their flight performance (with an increase in frequency).

Lateral force, lift force, and thrust

Without going into the details of the various parallelograms of forces – these are explained, for instance, in "The Evolution of Flight" (see below): the drag F_W of the wing generates a lateral force F_A that acts perpendicular to the inflow direction (somewhat missleadingly known as "uplift" even though this force does not necessarily point to the top) and thus also two beneficial forces: the lift force F_H and a forward-driving component F_V known as propulsion or thrust. Most wingbeat positions generate such beneficial components. During the hovering flight of hummingbirds, the thrust component disappears.

Demo video
http://tethys.uni-ak.ac.at/cross-science/flying-machine.mp4

Leonardo's Dream of Flying

Birds or bats as models?

Leonardo's flying machine is world-famous. Already as a child, he observed how birds spread their wings while flying upward and fold them again when moving down. Although the design of its flying machine is strongly reminiscent of bats, he called his machine the "Great Bird". In the two videos cited below, you can, in fact, recognise some parallels with the flight of vultures.

Can the apparatus fly?

Leonardo knew a lot about drag and the lift forces of propellers. His flying machine was supposed to be set in motion by human muscle power, but this would have been very difficult to accomplish. Its system of ropes and pulleys, however, has proven to be correct. The flying machine might have kept a human in the air for a short period of time…

Demo videos
http://tethys.uni-ak.ac.at/cross-science/vulture-flight.mp4
http://tethys.uni-ak.ac.at/cross-science/vulture-flight2.mp4

Demo video
http://tethys.uni-ak.ac.at/cross-science/flying-machine.mp4

Wing Twist

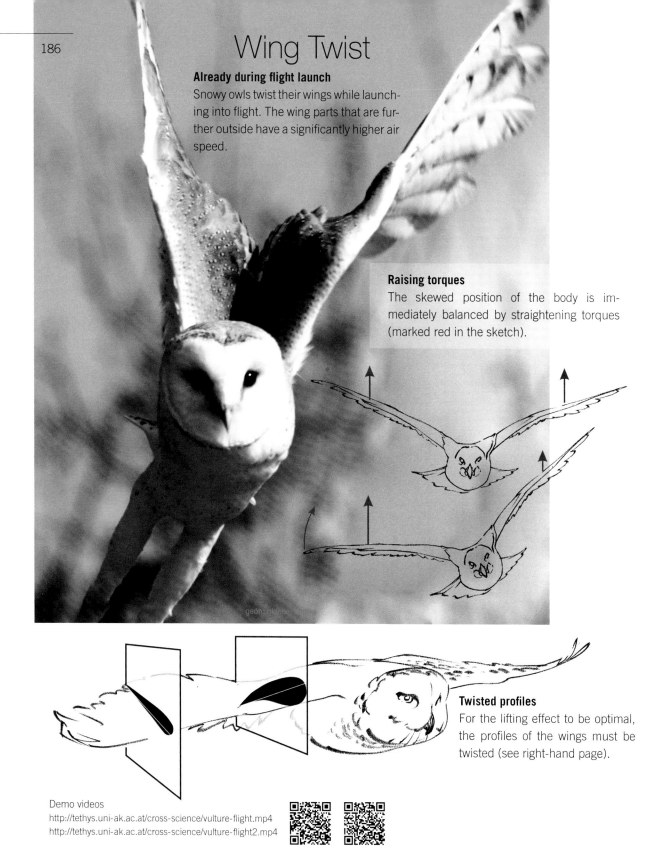

Already during flight launch
Snowy owls twist their wings while launching into flight. The wing parts that are further outside have a significantly higher air speed.

Raising torques
The skewed position of the body is immediately balanced by straightening torques (marked red in the sketch).

Twisted profiles
For the lifting effect to be optimal, the profiles of the wings must be twisted (see right-hand page).

Demo videos
http://tethys.uni-ak.ac.at/cross-science/vulture-flight.mp4
http://tethys.uni-ak.ac.at/cross-science/vulture-flight2.mp4

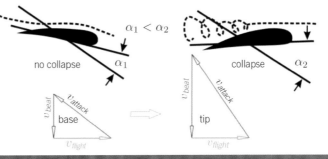

Inflow direction

Aerofoil-shaped wings generate optiomal lift for inflow angles α that are not too large. If α is too large, then the airflow will break down. α is obtained through vector addition: The body moves forward at velocity v_{flight} (green); in a certain position, the wing moves downward at velocity v_{beat} (blue). The sum will yield the inflow vector v_{attack} (red). The inflow angle is thus always inclined in the actual travel direction, and even more so, when the aerofoil profile is located at the outer edge of the wing.

Rotor blades of a wind turbine or helicopter...

...have a twisted or twistable profile for the aforementioned reasons. The helicopter in the images below is performing a spectacular manoeuvre. A complex mechanism (bottom right) enables it to keep the airflow from breaking down.

Demo video
http://tethys.uni-ak.ac.at/cross-science/flying-machine.mp4

Flying as If in Light Oil

The absolute size is crucial

It is by no means of no importance whether wings flapping in the air are big or small. A shrunken bird wing would be useless for small insects: for them air is a relatively viscous medium, comparable to light oil.

Flies beat their wings up to three hundred times a second. Bumblebees and honey bees beat them about two hundred times. Wasps still beat their wings more than a hundred times per second. The cockchafer pictured above has a frequency of 50 to 60 wingbeats per second. Dragonflies beat their wings almost 30 times per second and butterflies about 10 times.

Demo videos
http://tethys.uni-ak.ac.at/cross-science/bee-flying-off.mp4
http://tethys.uni-ak.ac.at/cross-science/wasp-compete.mp4
http://tethys.uni-ak.ac.at/cross-science/hornet-attack.mp4
http://tethys.uni-ak.ac.at/cross-science/fly-taking-off.mp4
http://tethys.uni-ak.ac.at/cross-science/cockchafer.mp4

Associations with ship propellers…

…are not far-fetched. Insects bend their wings depending on their flying speed and wingbeat frequency, but they are actually much more flexible. In any case, even with high wingbeat frequencies, wings do by no means move solely up and down, or back and forth during hovering flight. Experiments to prove this – with sufficient magnification and considerable decrease in wingbeat frequency – were in fact conducted in oil baths.

Measuring the frequency

The wingbeat frequency can be measured quite well with high-speed cameras. The image series of a fly at the top of the left-hand page was created with 1,000 shots per second. After about three images, we get a sort of dead centre, which suggests a frequency of about 300 wingbeats.

Demo videos
http://tethys.uni-ak.ac.at/cross-science/dragonfly-starting.mp4
http://tethys.uni-ak.ac.at/cross-science/dragonfly-landing.mp4
http://tethys.uni-ak.ac.at/cross-science/butterfly-like-in-oil.mp4

A minimum speed is needed for flight

Airplanes reach this speed by accelerating on the runway, helicopters through rapid rotations, and animals through wingbeats, though the frequency will depend on the size of the animal: the bigger the animal, the lower the frequency – coupled with high absolute velocity at the tip of the wings. Small insects reach the necessary velocity through extremely high frequency (first video). Smaller birds manage to launch straight up into flight by exerting extremely high force (second video).

Jumping off makes launching into flight easier

Large aquatic birds often launch into flight by pushing themselves off the ground or water surface with strong kicks or splashes (third video and right-hand page at the top). Big insects, such as stag beetles (first video on the next page,) often jump with their hind legs up into the air before launching into flight. The important thing is that they generate airflow right at the beginning of the flight. These jumps thus support the first wingbeats.

Demo videos
http://tethys.uni-ak.ac.at/cross-science/wasp-like-helicopter.mp4
http://tethys.uni-ak.ac.at/cross-science/bee-eater.mp4
http://tethys.uni-ak.ac.at/cross-science/heron-jumping-off.mp4

Slowing down with lowered landing flaps
Swans are amongst the largest animals capable of flight. Above: a troublesome start. Below: using the water surface and wing position to slow down its mass steadily.

Demo videos
http://tethys.uni-ak.ac.at/cross-science/stagbeetle-flying-off.mp4
http://tethys.uni-ak.ac.at/cross-science/cormoran-takeoff.mp4

Distributions:
Attraction and Repulsion

Filtering Nutrients

Animals and not plants

Even if it often does not look like it: most underwater organisms that look like ferns or flowers are actually animals (even when they have grown attached to the ground): plants have the exclusive capacity to perform photosynthesis – and for this they need sunlight, which is only available in the upper water layers. The resulting organic residues trickle towards the bottom of the sea, are carried around by currents. and feed a multitude of animals. Corals allow small polyps to spread over their skeletons. In a magnified close-up (image on the left-hand page), we can see that these polyps are equipped with feathered tentacles. The bottom line: the better the distribution of the polyps, the more efficient the filtering of nutrients. The demo video shows a possible computer simulation of coral growth.

Demo video
http://tethys.uni-ak.ac.at/cross-science/nearest-circle2d.mp4

Uniform Point Distribution

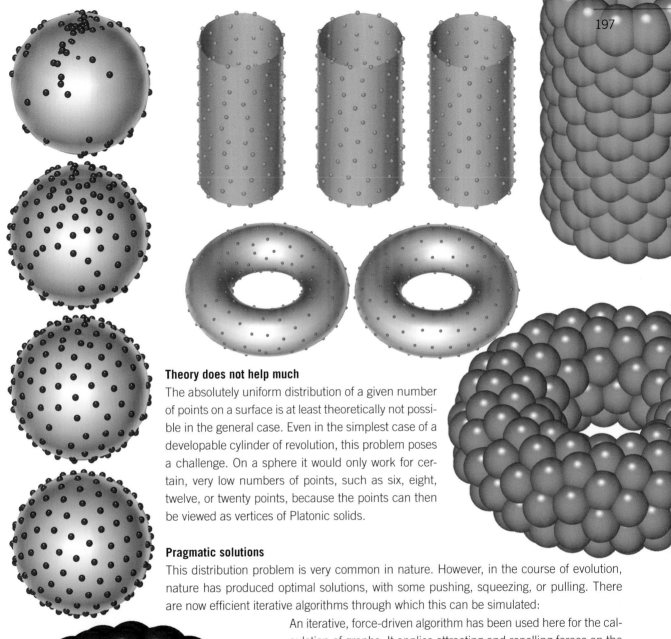

Theory does not help much

The absolutely uniform distribution of a given number of points on a surface is at least theoretically not possible in the general case. Even in the simplest case of a developable cylinder of revolution, this problem poses a challenge. On a sphere it would only work for certain, very low numbers of points, such as six, eight, twelve, or twenty points, because the points can then be viewed as vertices of Platonic solids.

Pragmatic solutions

This distribution problem is very common in nature. However, in the course of evolution, nature has produced optimal solutions, with some pushing, squeezing, or pulling. There are now efficient iterative algorithms through which this can be simulated:

An iterative, force-driven algorithm has been used here for the calculation of graphs. It applies attracting and repelling forces on the nodes (points) and is described in more detail on p. 132. With each iteration the distribution of points improves, and sometimes, after hundreds of improvements, the algorithm will arrive at the best solution. This is especially well illustrated in the image series on the left, where points are uniformly distributed on the sphere (first video).

Demo videos
http://tethys.uni-ak.ac.at/cross-science/points-on-sphere.mp4
http://tethys.uni-ak.ac.at/cross-science/points-on-cyl-and-torus.mp4

Voronoi Diagrams

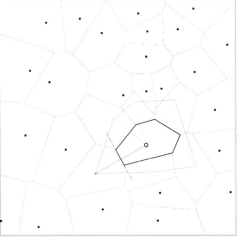

Derived from fairly uniformly distributed centres

In nature you can find numerous examples of plane or curved surfaces that are covered with patterns consisting of convex polygons (Voronoi diagrams). Such patterns can be generated through centres that are more or less uniformly distributed on a surface (see sketch on the left): the perpendicular bisectors of a centre to the neighbouring centres form the edges of the individual polygons.

Physical interpretation

If you think of the centres as points with equally strong gravity, then the respective convex polygon is the locus of all points that are attracted more strongly by the corresponding centre than by the other centres. That is why mud typically dries in a shape that resembles such a diagram: there are loci that dry faster than others and thus produce tensions that create faultlines in the middle.

The structure that forms a Voronoi diagram is not only especially flexible but also boasts high stability. This is probably the reason why the structures in leaves, the veins in dragonfly wings, and – as in the bottom left image – the underside of mushrooms offer fairly good approximations of such diagrams.

Iteration of the process

When a Voronoi diagram is created from a "point cluster", it can be improved both visually and physically by replacing the original centres with the plane centroids of the polygons.

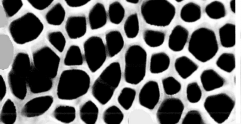

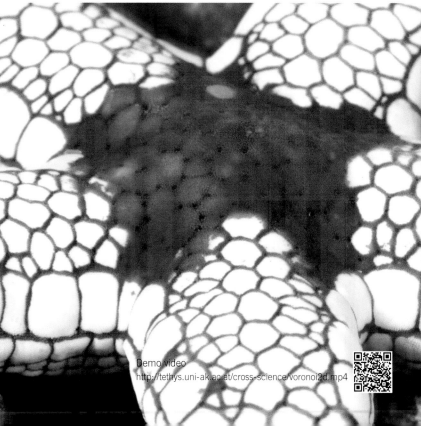

Demo video
http://tethys.uni-ak.ac.at/cross-science/voronoi2d.mp4

Creating patterns for camouflage

Some animals – especially aquatic animals – create patterns on their skin that are particularly reminiscent of Voronoi diagrams. This can be seen very well with the starfish on the left-hand page, but also with the longhorn cowfish at the bottom right, which belongs to the family of boxfishes and is equipped with a strong skin poison. The body of boxfishes is protected by a bone armour. The juvenile form is yellow (bottom left) and has "only" spots – the centres of the polygons.

In the case of starfishes and the turkana jewel cichlid at the top of p. 196, the points or patterns clearly serve as camouflage (the red colour is also a camouflage colour at water depths greater than ten metres). The yellow colour of young boxfishes is probably a warning colour.

Demo video
http://tethys.uni-ak.ac.at/cross-science/cowfish.mp4

Diagrams on a Sphere

Spherical polygons

Voronoi diagrams can be created on a sphere in a manner very similar to when they are created on a plane. Instead of straight-lined sides, we get great-circle arcs when intersecting the symmetry planes of two centres.

Optimisation of the distribution

Choose an arbitrary number of centres on the sphere and calculate the corresponding diagrams.

In the general case, this will look very irregular (blue sphere on the bottom left). Then we replace the centres with the spherical centroids of the correspsonding cells and determine once again the Voronoi diagram. This process can be repeated any number of times (blue spheres further right), with the diagram becoming increasingly balanced. Eventually, we will get balanced diagrams like the ones in the large image.

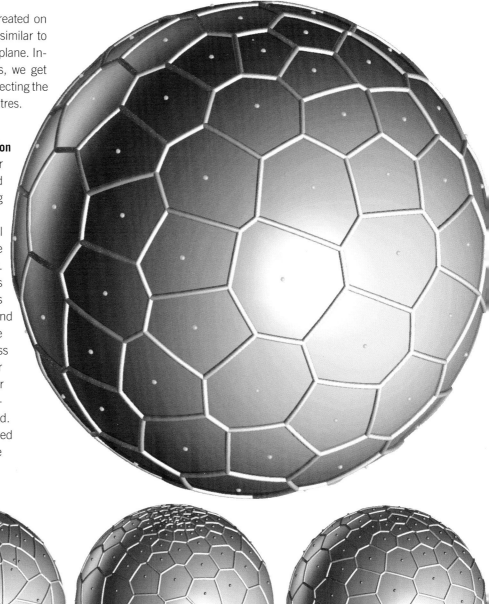

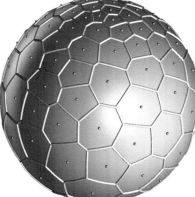

Demo video
http://tethys.uni-ak.ac.at/cross-science/voronoi-on-sphere.mp4

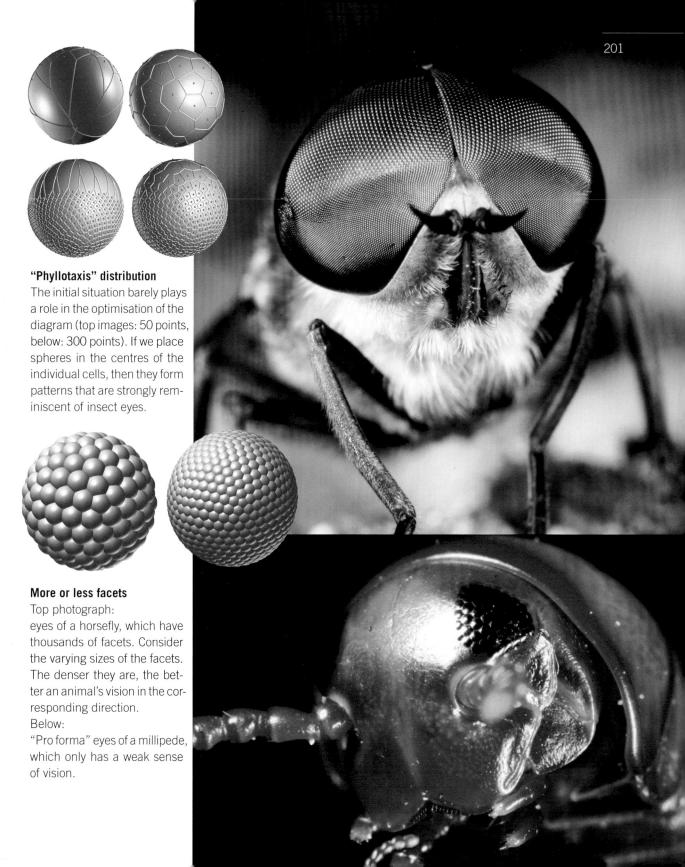

"Phyllotaxis" distribution

The initial situation barely plays a role in the optimisation of the diagram (top images: 50 points, below: 300 points). If we place spheres in the centres of the individual cells, then they form patterns that are strongly reminiscent of insect eyes.

More or less facets

Top photograph:
eyes of a horsefly, which have thousands of facets. Consider the varying sizes of the facets. The denser they are, the better an animal's vision in the corresponding direction.
Below:
"Pro forma" eyes of a millipede, which only has a weak sense of vision.

A Net of Hexagons on a Sphere

Compound eyes and wasp nests…

With compound eyes, we already encountered the following problem: can a double-curved surface like a sphere be covered entirely with regular hexagons?

The answer was: theoretically no. Nature, however, is more pragmatic: do these hexagons really have to be hexagonal, and must all polygons be hexagons?

A pragmatic approach

Depending on how close-meshed the sphere (or semi-sphere) is supposed to be tessellated, you could take the following approach: find a more or less uniform distribution of given points (this can be done via formulas or with the aforementioned algorithms) and then determine the corresponding spherical Voronoi diagram. This has been done in the three images on the right-hand page (third series). In order to create different figures and geometrical approaches, like on the middle left and bottom, you can do the following as on the top of the right-hand page: through inversion you can create a circle pattern in the equatorial plane of the sphere and project it stereographically from the north pole to the sphere, which results in a circle pattern on the sphere. If you enlarge the circles appropriately, they will intersect each other in fairly regular hexagons. However, you can also – as on the right-hand page in the middle – cover an icosahedron with a hexagon net and then project it onto the sphere.

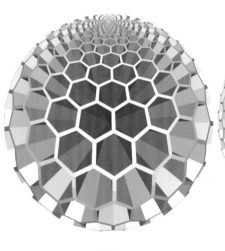

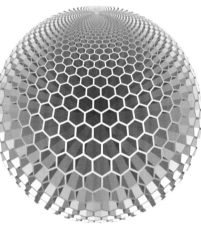

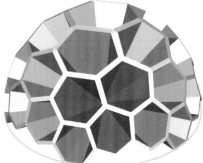

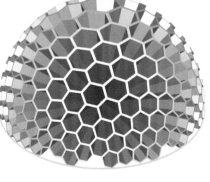

Demo videos
http://tethys.uni-ak.ac.at/cross-science/hexagons-on-sphere1.mp4
http://tethys.uni-ak.ac.at/cross-science/hexagons-on-sphere2.mp4
http://tethys.uni-ak.ac.at/cross-science/hexagons-on-tetrahedron.mp4
http://tethys.uni-ak.ac.at/cross-science/hexagons-on-icosahedron.mp4

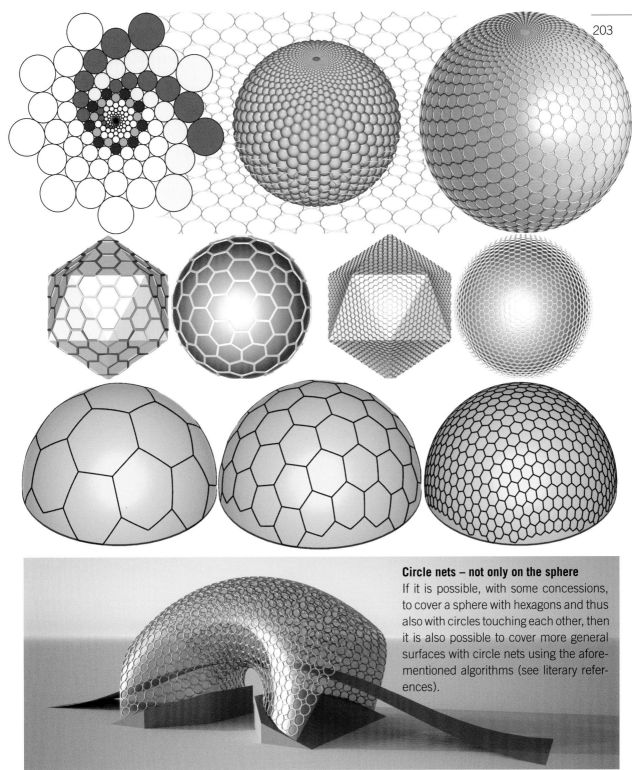

Circle nets – not only on the sphere

If it is possible, with some concessions, to cover a sphere with hexagons and thus also with circles touching each other, then it is also possible to cover more general surfaces with circle nets using the afore-mentioned algorithms (see literary references).

F. Gruber, G. Wallner, G. Glaeser **Force directed near-orthogonal grid generation on surfaces** *J. Geom. Graphics 14/2, 135-145 (2010)*
https://www.heldermann-verlag.de/jgg/jgg14/j14h2grub.pdf

Magnetic Nets

Magnetic spheres and rods

Imagine a net-like structure materialised by small magnetic spheres from which four equally long rods – that we also imagine to be magnetic – extend (first video).

This mobile net can be adapted quite well to common shapes by pulling the net little by little, so that it always assumes the desired final position. We must take into account that the net will obviously change as a whole and local changes thus have global impacts. As an illustration of this, a net has been cast over two or three spheres here.

In the second video, you can reasonably see how to go about this. Let such a net drop down, where it falls on a graph of function. Upon the first touch at the highest point, the net must be adapted. At the point of contact, the first sphere stops temporarily, while the other spheres continue to sink (and deviate from the perpendicular path in the process) until the graph is reached. At the end, the net must be pulled up slightly so the rods do not collide with the graph.

Demo videos
http://tethys.uni-ak.ac.at/cross-science/magnetic-sticks.mp4
http://tethys.uni-ak.ac.at/cross-science/falling-net.mp4

The benefits of this method

The image series in the middle shows how a sphere sinks into a magnetic net and pulls the net with it (the function graph here is partially formed by the lower semi-sphere).

We have thus found a way to approximate free-form surfaces through magnetic nets. Sets of four rods form "skew rhombi" which can be divided into two triangles through one of the two diagonals.

Elastic nets

However, the edges will then get out of control. Now we think about what happens if we define, for instance, the four vertices of a rectangular magnetic net in their final position. We can then assign a variable, pull-dependent length to the rods in order to simulate elastic nets, as has been done in the image series on the right. By colouring the rods, we can visualise quite well where the strongest pulling forces occur.

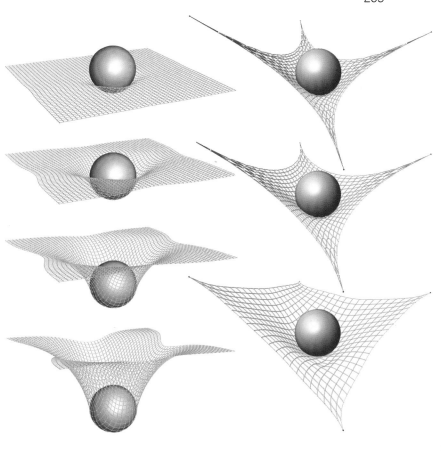

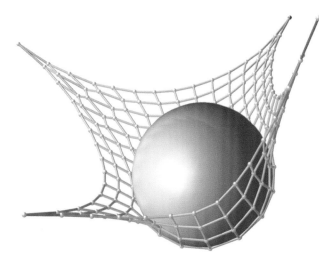

Demo videos
http://tethys.uni-ak.ac.at/cross-science/elastic-net1.mp4
http://tethys.uni-ak.ac.at/cross-science/elastic-net2.mp4
http://tethys.uni-ak.ac.at/cross-science/ball-in-net.mp4

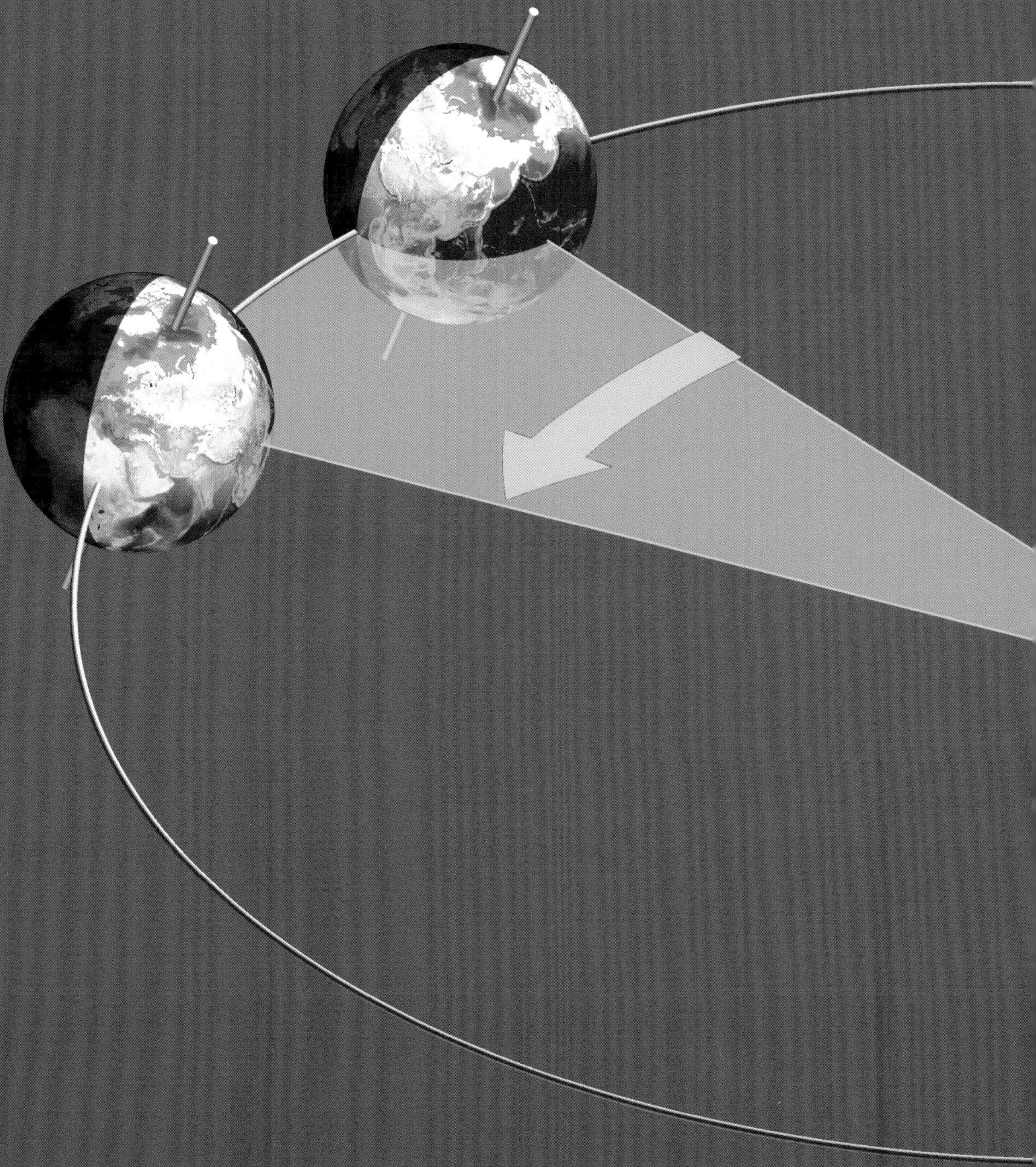

Our Solar System: Free Play of Forces

Spinning Motions

Spinning motions play an important role in space (p. 210). We want to look at the parameters that are relevant to this type of motion.

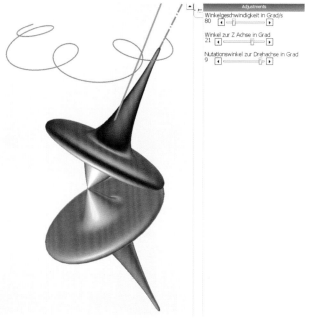

Simulation of a spinning top

For the sake of simplicity, we will assume that a spinning top moves approximately like a flat cylinder that is fixed to the ground by an axis passing through its centre. The centre of gravity lies in the cylinder's centre, or in the centre of the biggest bulge.

Self-rotation

A spinning top rotates comparatively fast around its own axis. If you determine its rotational velocity, you will immediately recognise that a high rotational velocity stabilises the object. In combination with the moment of inertia, the rotation will generate an angular momentum (represented here by the blue line) along the axis of rotation. In a scenario with no outside forces, the spinning top will rotate around its axis of rotation without hindrance.

Precession

However, in our model, gravity will always be acting vertically in the direction of the ground. So, without self-rotation, the spinning top would topple over. The rotation allows the spinning top not only to retain its inclination to the ground but also to rotate around an axis perpendicular to the ground. This type of motion is known as precession. Precession is the result of a torque generated by Earth's gravitational force. The torque's direction is perpendicular to the axis of rotation and the force of gravity. The torque can only move in this direction. Mathematically speaking, this is also a vector product. The velocity of precession is directly dependent on the strength of the acting force, the rotational velocity around the axis of rotation and the body's moment of inertia. The faster the spinning top moves, the slower is its precession.

Nutation

When you spin a top, you will often see it move in a wobbly manner. This phenomenon is known as nutation. It is generated by forces that briefly act against the axis of rotation, so that the axis no longer points in the direction of the torque. In order to preserve the torque, the spinning top must now make another compensatory motion. So, it starts to rotate its rotational axis around the direction of the torque. The velocity of nutation depends, on the one hand, on the degree of deviation and rotation around rotational axis and, on the other hand, on the shape of the spinning top and its moment of inertia. The faster its self-rotation and the greater the deviation, the faster will be the nutation. The spherical trajectory of a point on the spinning top's rotational axis is marked in red in the image.

A comparison with an actual experiment (see demo video) will show that these assumptions are very realistic.

Theory and demo videos

https://itp.uni-frankfurt.de/~luedde/Lecture/Mechanik/Intranet/Skript/Kap7/node5.html

http://tethys.uni-ak.ac.at/cross-science/kreisel.mp4

http://tethys.uni-ak.ac.at/cross-science/kreisel2.mp4

First Point of Aries

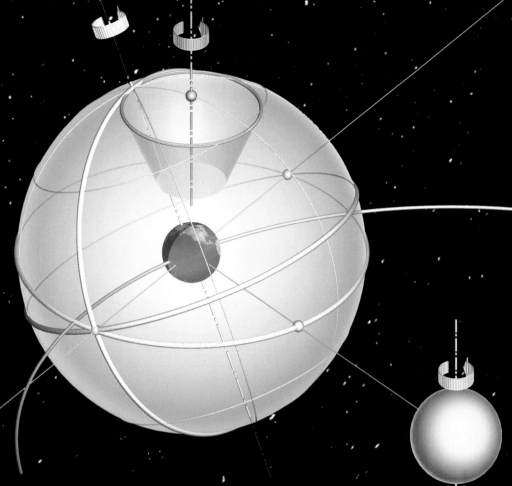

An important nodal point

The Earth's axis is currently tilted towards the ecliptic (the orbital plane of the Earth) by about $23, 4°$. There is a slight periodic variation in this inclination over the course of about $41\,000$ years (nutation). The equatorial plane of the Earth (equator marked in blue) lies at a right angle to the Earth's axis and intersects the ecliptic along a straight line (green). Since the axis remains nearly parallel over the course of many decades, so does the green line. The line points to a certain star constellation, and it does so for a long period of time. In our century, we call it the "Age of Aquarius".

Equinox

The green nodal line thus moves in a parallel direction along with the Earth and meets the Sun twice a year. Then, and only then, will the Earth's shadow – the dividing line between day and night – run through the poles, so that day and night are equally long. These moments in time define the beginning of spring and autumn.

Earth's precession

Over the course of almost $26\,000$ years, the Earth's axis will rotate around a normal line of the ecliptic. The nodal line will rotate along with the Earth's axis during this process, so that the spring and autumn equinox move once around the orbital ellipse.

Demo videos
http://tethys.uni-ak.ac.at/cross-science/precession-earth.mp4
http://tethys.uni-ak.ac.at/cross-science/change-of-equinox.mp4

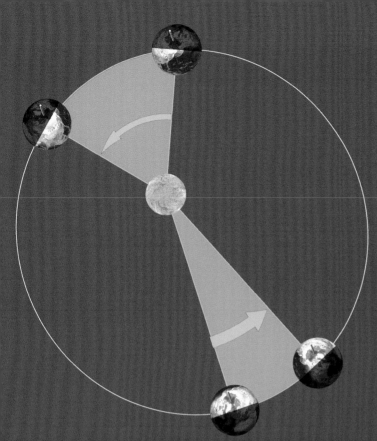

Kepler ellipses and free fall

The Earth is in free fall around the Sun, and in the course it moves on an elliptical orbit, with the Sun being one of the foci of the ellipse. According to Kepler's Second Law, equal areas of space are swept during equal time intervals (left image).

Northern winters are shorter than northern summers

The Earth currently reaches its perihelion, that is, the point closest to the Sun, in early January. As a result, the winter semester on the Northern Hemisphere is five days shorter than the summer semester. In about 13 000 years, the tables will have turned due to the Earth's precession, and the spring equinox will occur during the current autumn season. Then the Southern Hemisphere will have longer summers than the Northern Hemisphere.

When exactly do the seasons start?

According to the laws of geometry ([1]), we can determine the beginning of the seasons as follows: The (green) line that is orthogonal to the direction of the Earth's axis and passes through the Sun will yield the beginning of spring and autumn (the equinoxes), and the (blue) perpendicular line in the ecliptic will yield the beginning of summer and winter. The moments in time can be defined to the exact minute. The green line and the blue line are *not* the axes of the ellipse.

Demo videos
http://tethys.uni-ak.ac.at/cross-science/earth-around-sun.mp4
http://tethys.uni-ak.ac.at/cross-science/start-of-the-seasons.mp4

The Double Planet

Demo videos
http://tethys.uni-ak.ac.at/cross-science/double-planet1.mp4
http://tethys.uni-ak.ac.at/cross-science/double-planet2.mp4

A non-trivial mutual orbit

Earth and Moon "dance" around their joint centre of mass. Since the mass of Earth is 81 times greater than that of the Moon, this centre lies underneath the Earth's surface (its distance varies because the Moon's orbit is an ellipse). The Earth's centre is therefore not fixed in place but wanders, in relative terms, on a small ellipse (image series on the left-hand page).

It is not a rotation

The two celestial bodies are not – as in the top right image – chained together but connected only by gravity, though their axis of rotation remains stable. This motion is not a rotation in the classical sense (as with the two children in the image below, who are both rotating around an axis of rotation), becauses then almost all points on Earth would have a different orbital velocity. Instead, all points on and inside the terrestrial globe have the *same* velocity – namely that of the Earth's centre – because the direction of the Earth's axis remains unchanged (see also p. 210).

Lunar phases and eclipses

The Moon's orbit around the Earth in about four weeks creates the lunar phases (in the image series on the left-hand page: waxing moon, half moon, nearly full moon, below that a potential lunar eclipse during full moon, waning moon. Since the double planet orbits around the Sun by about thirty degrees within four weeks, it will take two days longer to reach the same lunar phase in the next month.

"Earthshine"

Due to the reflection of the Sun's light off the Earth's surface, the Moon is not entirely dark during the days of the new moon (the two bottom photographs show such a situation – the photograph at the very bottom was heavily overexposed, which has resulted in almost the same picture as during full moon – but with very little light).

The same Moon as above but overexposed by seven f-stops. The result is "a near full moon". Leonardo da Vinci already knew that that was caused by the reflection of sunshine off the Earth.

The Tides

The tidal force fields of the Moon and the Sun add up. If the new moon occurs between the Earth and the Sun, it can result in an especially high spring tide.

During half moon, the directions to the Moon and Sun are perpendicular, and the ebb and flow of the tides is more moderate (neap tide).

During full moon as well, when Moon and Sun are positioned opposite each other, we get a – somewhat weaker – neap tide.

Demo video
http://tethys.uni-ak.ac.at/cross-science/tides.mp4

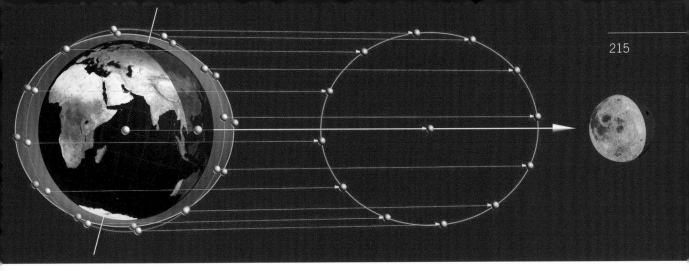

The deformation of the Earth's surface through the Moon

Gravitational forces act on the entire planet, but tidal forces can best be observed on the oceans.

Let us focus on the Moon for now and look at the image above: the water mass is drawn in the direction of the Moon – by a few percent more on the side of the Earth that is facing the Moon than on the side that is facing away from the Moon. Exaggerated strongly, this results in an oval marked in red here that resembles an ellipse (viewed in three-dimensional space, we are dealing with an ellipsoid).

Gravity is balanced entirely by the centrifugal force at the centre of the Earth – after all, the double planet is in balance. Since – due to the constant direction of the Earth's axis – every point is affected by the same centrifugal force as the Earth's centre, the red oval is centred around the Earth's centre.

Ebb and flow twice a day

This results in an ocean surface that is bulging on two sides: the maximum distances are found at the points that are closest to Moon and furthest from the Moon. Now, within about 24 hours, the Earth rotates around its own axis and the Moon orbits around the Earth by roughly 10–15 degrees (the angular velocity is not constant). As a result, ebb and flow occur *twice* within almost 25 hours.

The Sun is also in on the act

The Sun also creates such an oval, though one that is only about half as pronounced as the tidal force field of the Moon – since we are so far away from the Sun, the differences in the distances of the points on the Earth are much smaller percentage-wise. If we add up the two ovals, we get a new oval that is particularly pronounced during new moon and full moon.

Tidal force fields of the Moon (left), the Sun (centre), and the superimposition of both (right).

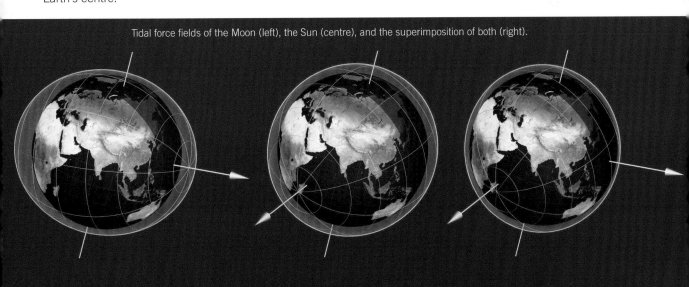

Solar and Lunar Eclipse

Solar eclipse: a rare event

On the Earth's sky, the Sun and the Moon coincidentally have almost the same diameter of half a degree. There can only be a solar eclipse when the new moon covers the Sun almost entirely. In addition, the Moon's umbra has a relatively small diameter, so that solar eclipses are always limited locally.

Beginning solar eclipse (top left space situation).

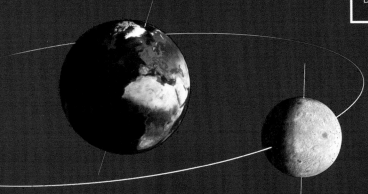

The Moon's orbital plane is tilted

The "main problem", however, is that the Moon's orbit is tilted by about five degrees relative to the orbital plane of the Earth. Yet, the Moon's orbital plane rotates over the years, and with it the intersection line with the Earth's orbital plane. Only if the new moon happens to be positioned on this intersection line, a solar eclipse occurs.

Seemingly variable diameter of the Moon

The Moon's diameter varies quite significantly due to its elliptical orbit. It is thus possible for the new moon to appear especially big (by analogy with the super full moon, we could call it a super new moon). The Moon then easily manages to cover the Sun entirely. With a smaller new moon, the result is ideally e ring-shaped eclipse (in the picture at the bottom, the umbra is cast over Egypt).

Annular solar eclipse (bottom left space situation).

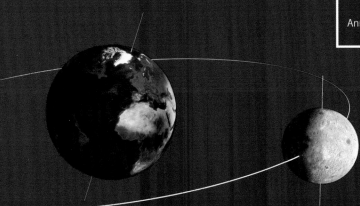

Only a few minutes

Due to the Earth's self-rotation, the umbra rushes over the Earth at a high speed (usually over 1500 km/h). A total solar eclipse, therefore, lasts only briefly.

Demo videos
http://tethys.uni-ak.ac.at/cross-science/solar-eclipse.mp4
http://tethys.uni-ak.ac.at/cross-science/orbit-of-the-moon.mp4

Beginning lunar eclipse (top right space situation).

Lunar eclipse: a more common event

During a lunar eclipse, the Earth casts a shadow onto the full moon. The Earth's diameter is almost four times as wide as that of the Moon, so that lunar eclipses occur much more often and they are visible from all points of the Earth that are facing the Moon. Lunar eclipses are less spectacular though (as the full moon also tends to be veiled by clouds).

Blood Moon (bottom right space situation).

Lunar eclipses last much longer

The shadow of the Earth can be sufficient to darken the full moon for up to an hour and a half. Remarkably, while the Moon is within the Earth's umbra, it can turn a reddish hue ("Blood Moon"), which is caused by the refracton of light on the Earth's atmosphere.

Demo video
http://tethys.uni-ak.ac.at/cross-science/lunar-eclipse.mp4

A Few More Things about the Moon

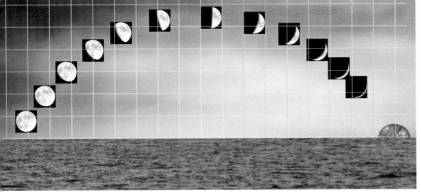

Where is the Moon during sunset?

The Moon is delayed each day by about fifty minutes on average. During new moon, it appears close to the Sun. Then the crescent moon waxes until full moon, when the Moon eventually rises on the opposite side of the Sun. While it is waxing, the crescent moon appears before sunset. The waning moon, on the other hand, is only visible after sunset.

The inclination of the crescent moon

The Moon's terminator (the self-shadow boundary) always runs almost exactly through the lunar poles. It appears as a more or less bulbous ellipse, with the focus of the major axis indicating the Moon's axis quite accurately. Due to the Earth's rotation, this major axis rotates by about 15 degrees per second, and it does so in clockwise direction if we look to the south, and otherwise in counter-clockwise direction (see the first video: the direction of rotation changes when we look in or against the direction of the Earth's axis).

How high does the Moon rise?

The new moon, which appears close to the Sun, rises as high as the Sun, that is, high in the summer and low during winter. The full moon, on the other hand, behaves like the Sun if it was shifted by half a year: it rises high in winter but not in summer. The half moon lies somewhere in between, behaving almost like the Sun during the equinoxes. It reaches the highest point not at noon or midnight but instead almost exactly in the middle. In any case the Moon travels – like the Sun and also the stars – from east to west due to the rotation of the whole Earth globe.

Demo videos
http://tethys.uni-ak.ac.at/cross-science/orientation-of-rotation.mp4
http://tethys.uni-ak.ac.at/cross-science/moon-views.mp4
http://tethys.uni-ak.ac.at/cross-science/moon-phases.mp4

The Moon is actually a helpful light source

With full moon and a cloudless sky, it will not get dark all night. During half moon, either the first half of the night (waxing moon) or the second (waning moon) will be sufficiently illuminated.

...Sun and Moon
2...Sun, no Moon
1...no Sun, but Moon
0...neither Sun nor Moon

Polar nights are by no means pitch-black throughout

During winters in the high North (and obviously half a year later close to the South Pole), the Moon assumes an important role as the sole light source! This applies especially to the lunar phases from half moon over full moon to the next half moon, so in sum to half of all polar nights! If there is snow and ice, you only need relatively little light to recognise your surroundings clearly. The large image here shows how polar nights at the North Pole are illuminated by the Moon (not a full moon here because then it would appear exactly opposite the Sun). In the first video, you can see – as an animation – the corresponding situation close to the South Pole. The second video shows what the full moon is like during polar nights in the south: it remains above the horizon for 24 hours and travels due to the Earth's rotation in the opposite direction to the northern hemisphere, from right to left.

Demo videos
http://tethys.uni-ak.ac.at/cross-science/antarctic-in-winter.mp4
http://tethys.uni-ak.ac.at/cross-science/full-moon-antarctic.mp4

A Star and Some Matter around It

Now there are only eight planets…

…moving around our Sun, which has a mass that is almost a thousand times greater than that of all planets together. Everything is in motion! Even the Sun rotates around an axis, and its centre "wobbles" a little. The planets move, in accordance with Kepler's First Law, in elliptical orbits and, in accordance with Kepler's Second Law, with constantly varying speed.

Nothing in this system is set in stone

None of the relative motions is a rotation, except for the celestial bodies' self-rotation, and even the axes of these self-rotations are subject to twists and fluctuations.

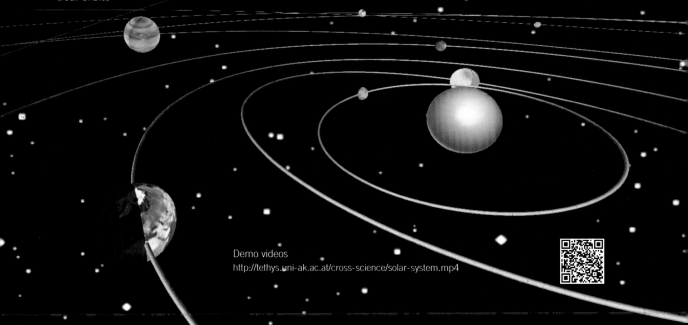

Demo videos
http://tethys.uni-ak.ac.at/cross-science/solar-system.mp4

Size ratios

We are used to images of our solar system where you "can make out something". The actual size ratios must be ignored for these images. Even Jupiter, which has a diameter that is ten times wider than that of Earth – and thus a thousand times the Earth's volume – is diameter-wiser a decimal power smaller than the Sun – and smaller by three decimal powers when it comes to volume.

The image series to the right shows the following things from top to bottom

● the actual proportions,

● the planets magnified 1,600 times, and the Sun without magnification,

● a "close-up" of the inner planets 80 times magnified in relation to the Sun,

● below the same scene but with Jupiter and Saturn,

● a comparable scene but with the Sun "only thirty times" less scaled than the planets.

Only Venus steps out of line

All planets orbit in the same direction around the Sun, and all planets, with the exception of Venus and the Sun, rotate around their individual axis in the same direction. Apparently, Venus was involved in a massive collision with another primordial planet, which then led to this "mishap".

Demo videos
http://tethys.uni-ak.ac.at/cross-science/solar-system2.mp4
http://tethys.uni-ak.ac.at/cross-science/rotation-speeds.mp4
http://tethys.uni-ak.ac.at/cross-science/as-seen-from-mars.mp4

Illusions:
Fake or Real?

Geometry Can Explain a Lot

On the previous double-page…

you can see two original photographs that will propably leave you wondering: What does the photograph on the left show? Is the elephant photograph a photomontage? In times of artificial intelligence, one should be very sceptical. However, there is a natural explanation for both images.

Cat eyes

The eyes of some animals reflect back when they are illuminated. This is due to reflectors inside their eyes. Due to strong refraction on the retina (bottom right image), light can enter and exit nearly sideways.

So, the eyes will glow when they are not illuminated from the front. This even works without refraction, if we use mirrored cube corners (first video): whichever way we shine a light into the eyes, the light will return to the light source – a simple but brilliant discovery, which underlies the design of bicycle reflectors (second video). The top right image shows a detailed view of the image on p. 222 (with a caption): the visible mirrored surfaces reflect the bordering edges. It takes quite a long time "to understand" this photograph.

Demo video
http://tethys.uni-ak.ac.at/cross-science/reflecting-corner.mp4

The trumpeting elephant

Usually, you use a telephoto lens to photograph objects that are far away and often also quite big. However, you can also use it to make distant, small objects appear big. On the other hand, a trumpeting elephant bull standing nearby will fit neatly into the image if we photograph it using an ultra-wide-angle lens. Of course, this trumpeting elephant could also just be a well-made fist-sized model of the actual animal. The model might then look almost as imposing as in reality. In order to adapt the surroundings to the size of the model, the following trick has been applied on p. 222: the trees that are visible in the background (and about a hundred times bigger than the elephant model) are about 30-times further away than one might assume, namely over 100 m. They thus appear small enough to "fit into the overall image". The horizontal mirror that has been positioned in the landscape mimics a water surface.

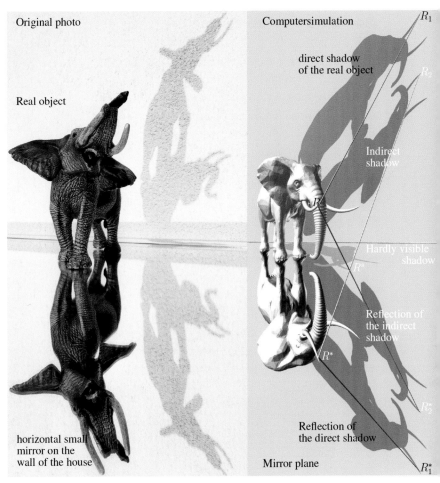

Original photo

Computersimulation

R_1

direct shadow
of the real object

R_2

Real object

Indirect
shadow

R

Hardly visible
shadow

R^s

Reflection of
the indirect
shadow

R^*

Reflection of
the direct shadow

R_2^*

horizontal small
mirror on the
wall of the house

Mirror plane

R_1^*

How many elephants are in the image?

The top centre image was created with the same equipment (animal model and mirror) on a wall. To explain this effect, we need to rely on our geometrical knowledge of perspectives, shadows, and reflections. The model elephant (represented by point R, for which the striking tip of the right tusk has been chosen) has a mirror image in the horizontal plane (point R^*). From the four recognisable shadows on the perpendicular wall, only two actually count – the other two are reflections of the first two.

The Sun illuminates each point on the wall and thus does not cast any direct shadows there. The reflection creates a mirrored Sun, which illuminates the scene *from below*. In those areas of the wall that are not illuminated by the mirrored Sun, we get the visible half-shadows, which do not differ in their intensity.

First we get a direct half-shadow on the wall (represented by point R_1, mirror image R_1^*). The shadow of the real object (point R^s) through the (real) Sun in the mirror plane can only be seen in the computer simulation, but not in the photograph! This non-illuminated spot on the mirror creates the other half-shadow (point R_2, mirrored point R_2^*) in the reflection on the vertical wall.

Confusion through Multiple Mirror Reflections

What happens in a mirror reflection?

Geometrically speaking, one can understand the following line of thought quite well when looking at the reflection of an object on a plane mirror: The mirror plane generates a "virtual counter-world" that is partially visible through the mirror frame and partially not. The virtual counter-world is mirror-inverted in the sense that the coordinate perpendicular to the mirror plane is reversed.

Two mirrors perpendicular to each other

In a next step, we will counsider double mirror reflections where the two mirror planes are perpendicular to each other (top image to the right). Now the following happens: each mirror generates a virtual counter-world that is partially visible through the corresponding mirror frame. These counter-worlds are simultaneously so "real" that each of them is mirrored again by the two mirrors – and since they are thus doubly reversed, they no longer appear mirror-inverted. Due to the right angle between the mirrors, the double mirror reflections are identical. One

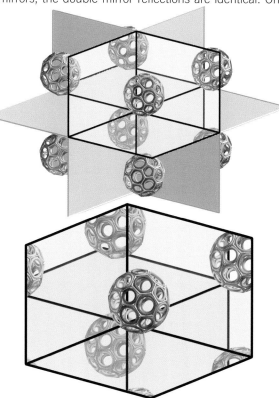

part is visible through the first mirror frame, and the complementary part is visible through the second frame.

Three pairwise perpendicular mirrors

In the middle right image, a third mirror has been added, creating a "mirrored corner". The horizontal mirror generates, altogether, four additional virtual worlds, which are partially visible through the third mirror frame. The number of mirrors determines if the counter-worlds are mirror-inverted or not. This is clearly illustrated by the top computer graphic on the left-hand side of this page. By cropping out those parts of the counter-world that are not visible in the mirror, we get photorealistic images that are no longer as easy to analyse.

Generalisation

When the two mirrors do not meet each other at a right angle but instead form an angle of, say, $60°$, as in the images to the right, more and more virtual worlds come into play (in this concrete case, we get, in addition to the two real spheres, ten virtual spheres that are partially visible through the mirror frames).

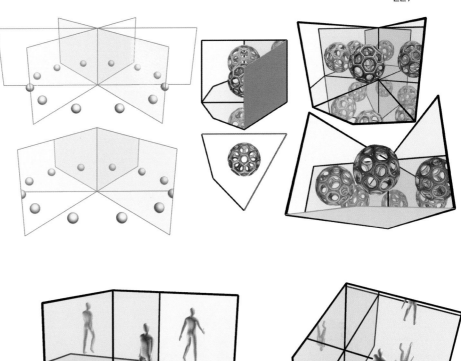

Representation in realtime

Multiple mirror reflections can be handled quite well through so-called "Raytracing" softwares, but image generation can take relatively long – unless you use special graphics cards.

However, with regular graphics cards, you can use special settings to add enough images per second (at leaast 20 images) to achieve a smooth representation of movement:

From the outset, you generate enough virtual objects and represent them in the correct order – and only those parts that are visible through the corresponding mirror frames. This is illustrated by the demo video with the human figure and the "buckyball", an object with many symmetries that is derived from an Archimedean solid.

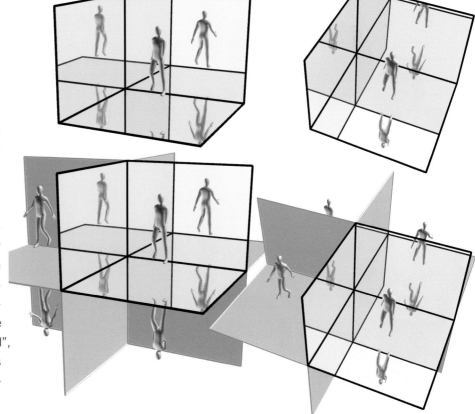

Demo videos
http://tethys.uni-ak.ac.at/cross-science/multiple-reflections.mp4
http://tethys.uni-ak.ac.at/cross-science/deception.mp4

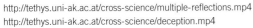

Mysterious Crop Circles

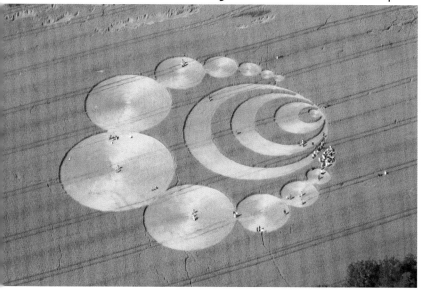

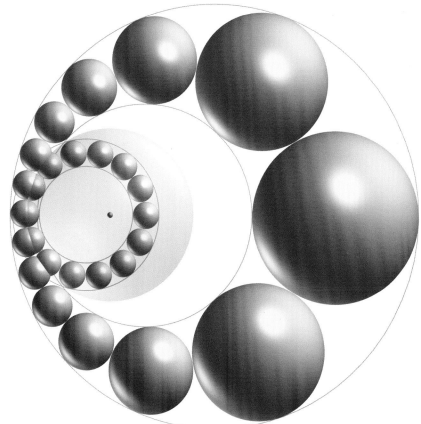

A beautiful circle configuration

By means of the so-called inversion on a circle, you can create beautiful and non-trivial figures as in the image below: we start with a given number of congruent circles touching each other and arrange them in circular manner such that their centres correspond to the vertices of a regular polygon. Now they fit into a torus. If we apply an inversion, then the two circles of the torus are transformed in the symmetry plane into two new circles. Spheres are thus not transformed into congruent spheres that touch each other and form a chain. However, the centres of these spheres lie on an ellipse. Without inversion, it would be difficult to create such a figure.

Aliens?

The image was created after the author had been shown photographs of crop circles like the one at the top left. Although it has been proven repeatedly how humans create this kind of hoax, there are still plenty of people who believe that these patterns are messages by "extraterrestrials".

Tractor trails

On the right-hand page, you can see screenshots from a simulation. This animation (see video) shows how this pattern can be efficiently created. There is a giveaway clue in the – ever-present – trails left by agricultural machines, because the people who produce these crop circles must move across the corn fields along such trails (photograph on the left of the right-hand page).

Reference and demo video
https://en.wikipedia.org/wiki/Crop_circle
http://tethys.uni-ak.ac.at/cross-science/crop-circles.mp4

The Zoetrope

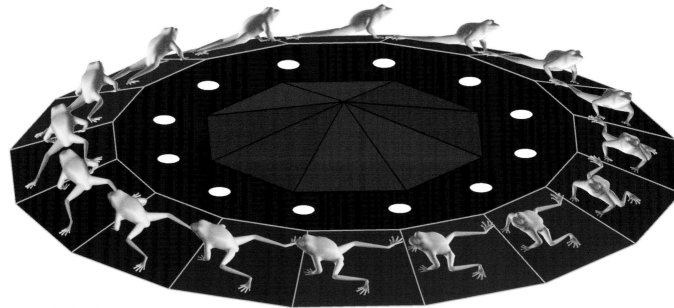

Illusion of motion

If you spin a disc with a series of uniformly distributed objects that represent intermediate stages of an object's movement, and if you spin it at a certain angular velocity, then the brain is tricked into seeing only one moving object. This pre-cinematic apparatus is known as zoetrope. The image above shows a computer model of a frog in 16 different positions. If you rotate the disc at an increasing angular velocity, then it appears as if the frogs rotate in clockwise or counterclockwise direction. From a certain number of revolutions onwards – when the positions of the frogs fall into line within less then a twentieth of a second – then you will see 16 frogs circling their legs on the disc.

The image below shows a pile of cubes that are one by one modelled in such a way that they perform individual rotations. With a sufficiently high image repetition rate, the pile itself is set in motion…

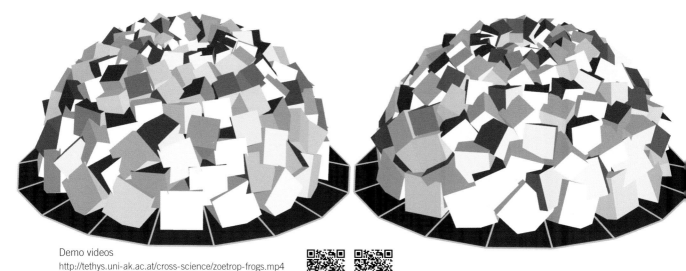

Demo videos
http://tethys.uni-ak.ac.at/cross-science/zoetrop-frogs.mp4
http://tethys.uni-ak.ac.at/cross-science/zoetrop-cubes.mp4

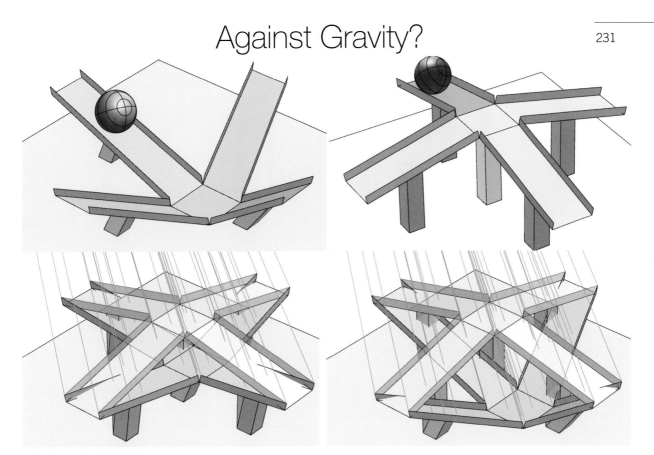

Our brain lets itself be deceived...

Kokichi Sugihara has spent some thoughts on how one can make spheres appear to be rolling upwards. To achieve this, you must force the viewpoint into a certain position. If from this position you think that you are looking at a stable and balanced construction, then your brain will immediately interpret the height of certain points in the construction, and it will tell you: the platform at the centre in the top right image has the highest position.

However, the construction is not symmetrical and stable. Its vertices lie somewhere on the projection beams through the image points of the construction. Among the inifinitely many possibilities, one should pick a non-symmetrical construction with a platform that lies lower. With more complicated objects, one will apply a spatial-perspectival collineation in order to calculate the corresponding object.

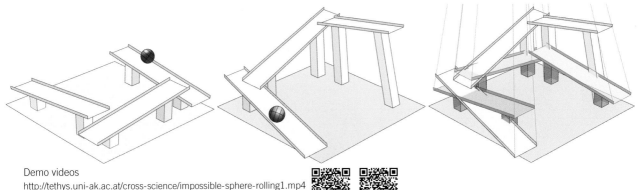

Demo videos
http://tethys.uni-ak.ac.at/cross-science/impossible-sphere-rolling1.mp4
http://tethys.uni-ak.ac.at/cross-science/impossible-sphere-rolling2.mp4

Various Other Illusions

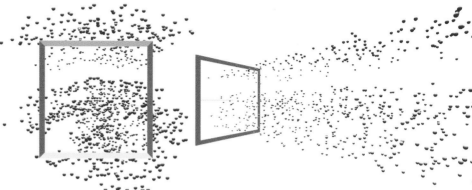

When you are tied to a certain position

When it is possible to position a person on a very specific spot, you can delude them into thinking quite a number of things. If you are forced, for instance, to look through a peephole at a supposed picture frame which shows a striking pixelated image like the one on the top left, you will believe that you are seeing something completely different than you would from other positions. If you move a bit closer (top centre), the pixels will somehow disappear. A view from a very different position reveals a secret: spheres of different sizes have been distributed across the space here (top right).

Star signs

We are in one such forced position when we look at the night sky and recognise various star signs, such as the Plough or Big Dipper, through which we can locate the North Star (Polaris, image series below). Here we see the stars of our Milky Way, which differ in brightness and are by no means equally far away.

Paper scraps falling from the ceiling

Imagine sitting in a lecture hall, enjoying some images projected onto a wall by a beamer or slide projector (top of the right-hand page).

Now someone decides to pull a silly joke and starts to let paper scraps flutter down from the ceiling into the light beam. People sitting on the sides of the auditorium will basically not see anything anymore. Viewers close to the projector, on the other hand, will barely notice a difference! A perfect illusion has occured here: the paper scraps are also illuminated by the projector. On each scrap, we get an image that is perspectively more or less extremely distorted and represents a part of the original

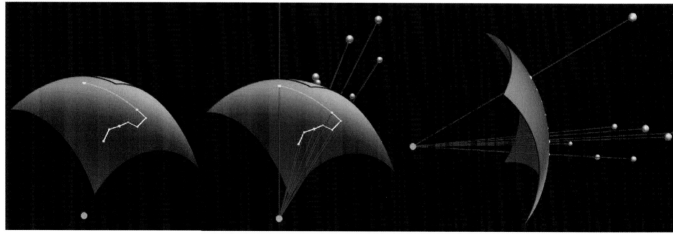

Demo videos
http://tethys.uni-ak.ac.at/cross-science/illusion-of-an-eye.mp4
http://tethys.uni-ak.ac.at/cross-science/big-dipper.mp4

image. Moreover, a drop shadow of the paper scrap is cast on the projection wall. Viewed from the lens centre, all this is not visible.

When there is only a skew polygon

"Thinky the Dragon" (image below, see also the cited website) appears to sit on a pedestal. The pedestal is actually a spatial cuboid.

However, Thinky is a polygon folded in different directions. When you rotate the scene slightly, it looks as if the dragon is rotating its head. It is only when you rotate further that you will recognise the underlying trick. Our brain has been trained to perceive objects as three-dimensional solid bodies. Even with an illustration video, you might fall for this trick time and time again.

Webpage and demo videos
https://www.thinkfun.com/teachers/thinky/
http://tethys.uni-ak.ac.at/cross-science/dragon-thinky.mp4
http://tethys.uni-ak.ac.at/cross-science/falling-leaves.mp4

Demo video
http://tethys.uni-ak.ac.at/cross-science/pendulum-system.mp4

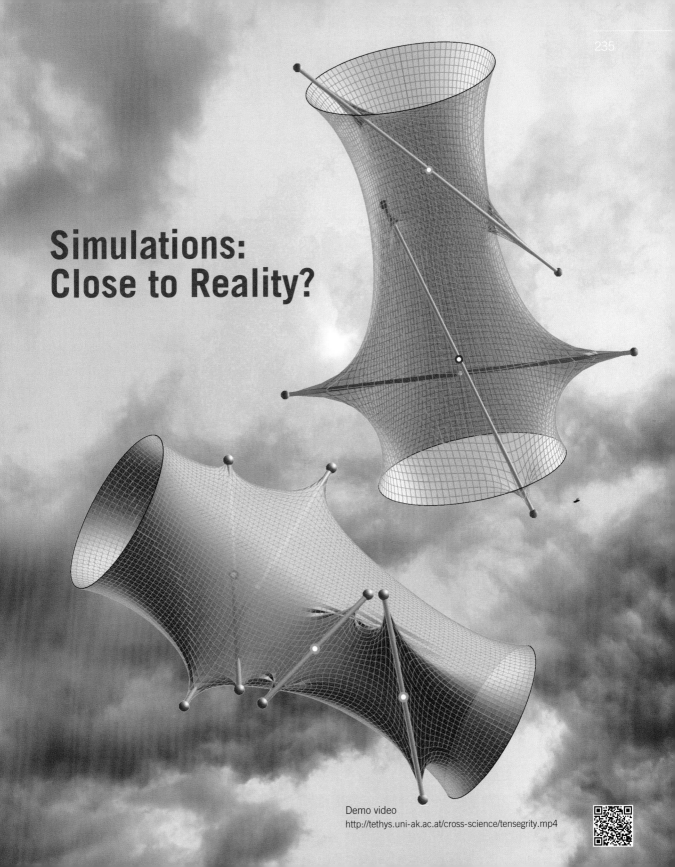

Simulations:
Close to Reality?

Demo video
http://tethys.uni-ak.ac.at/cross-science/tensegrity.mp4

Simulations in Theory and Practice

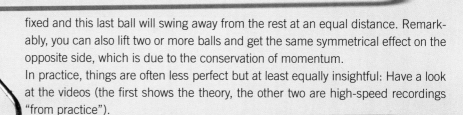

Swinging ball pendulum or Newton's cradle

This example is certainly well known to everyone from physics class: A set of balls hang from a rod, as pictured in the photograph, so that they are in contact when in a resting position. If we now lift one of the outer balls, keeping the string stretched, and let it drop back, then all balls except the outer one on the opposite end will remain fixed and this last ball will swing away from the rest at an equal distance. Remarkably, you can also lift two or more balls and get the same symmetrical effect on the opposite side, which is due to the conservation of momentum.

In practice, things are often less perfect but at least equally insightful: Have a look at the videos (the first shows the theory, the other two are high-speed recordings "from practice").

Demo videos
http://tethys.uni-ak.ac.at/cross-science/newton-simplified.mp4
http://tethys.uni-ak.ac.at/cross-science/newton1.mp4
http://tethys.uni-ak.ac.at/cross-science/newton2.mp4

The sphere in the bowl

The animations above show how a sphere rolls in differently shaped bowls. Minimal changes in the parameters will immediately impact the outcome (initial velocity, initial angle).

Demo videos
http://tethys.uni-ak.ac.at/cross-science/rolling-sphere.mp4
http://tethys.uni-ak.ac.at/cross-science/rolling-ball1.mp4
http://tethys.uni-ak.ac.at/cross-science/rolling-ball2.mp4
http://tethys.uni-ak.ac.at/cross-science/rolling-ball3.mp4

Gravitation is the driving force

The sphere accelerates in a manner that has been shown by Galilei in his famous experiment. However, the sphere is not rolling on a tilted plane but instead on a graph of function.

Swarm Behaviour

Demo videos
http://tethys.uni-ak.ac.at/cross-science/swarm-calm.mp4
http://tethys.uni-ak.ac.at/cross-science/swarm-stressed.mp4
http://tethys.uni-ak.ac.at/cross-science/sharks-hunting.mp4
http://tethys.uni-ak.ac.at/cross-science/sardine-run.mp4

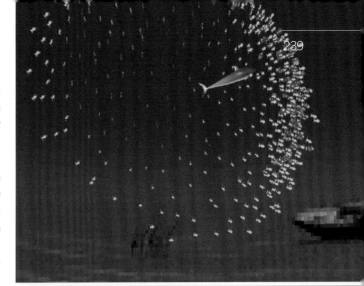

A perfect choreography?

Small fishes not rarely exhibit – like many birds – striking swarm behaviour. Interestingly enough, there is no leader that decides what should happen next. Instead, the whole choreography can be regarded as the product of many individual reactions.

Three rules

A fish in a swarm actually just wants to stay at a relatively close distance to its neighbour, e. g-. a distance of about a body length (rule 1). During the search for food, the individuals tend to move leisurely to where they can find more food (first video on the left-hand page, rule 2).

If a predator appears, the individuals in its vicinity will understandably flee from the site of danger (rule 3). Their neighbours might not have noticed the source of danger, but – after a short reaction time – they will want to rush after their fleeing neighbours immediately. Overall, this will trigger a chain reaction (shock wave) that can affect individuals that are far away from the danger (third and fourth video on the left-hand page). If the danger has passed, then the fastest escapees will try to rejoin the swarm in accordance with rule 1. After a certain time – and possibly until the next attack – order is once again restored.

The whole chaos is defined by these rules that can be programmed relatively easily (first video), though there are hardly any limits to the diversity of the swarm's shapes.

Simulation of a traffic jam

The image below and especially the second video on this page show how easily we can get an "accordion effect" when the distances between the cars are too close, so that the car driver's reaction time cannot keep up.

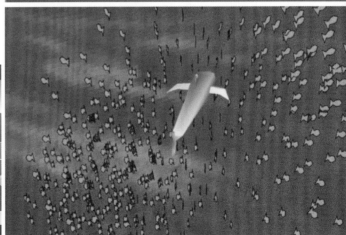

Demo videos
http://tethys.uni-ak.ac.at/cross-science/swarm-simulation.mp4
http://tethys.uni-ak.ac.at/cross-science/traffic-jam.mp4

Imitating Realistic Movements

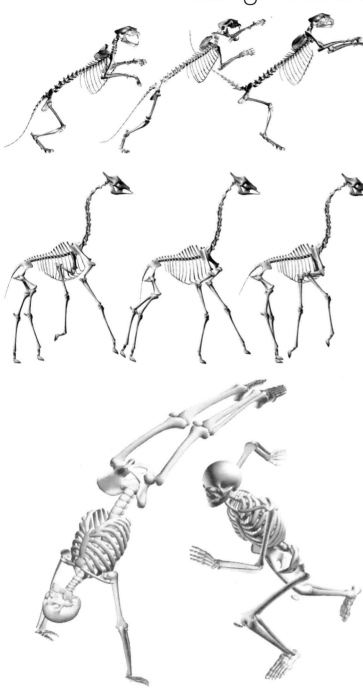

Over thousands of years of evolution…

…Nature has perfected the movements of animals: be it a small cat that already moves its body perfectly in order to catch a lure, or a deer that jumps across rough and uneven terrain at full speed (right-hand page). There is no doubt that Nature has already optimised motions. If evolution happens to find a better solution through mutation so that individual animals can move even more efficiently, then these animals will reproduce at a greater frequency and thus propagate their mutated genes. This raises the following question for programmers: how can such a perfect movement be simulated?

Analysis with reference images and videos

A possible solution to this problem is the following: first we gather reference images and videos, which are preferably precise frontal and lateral views of the subject. (That is why the lion sequence at the top right might be spectacular but it is not 100% ideal for this purpose. The bottom right series of the galloping giraffe, on the other hand, is much more suitable.)

Model of a skeleton

As a next step, we load the model of a skeleton in a professional software that provides modules for so-called inverse kinematics (e.g. the freely available modelling and animation software Blender). Ideally, the model should appear in a symmetrical rest pose. This is then followed by the "rigging process", which generates a bone-like structure with flexible joints that serves as a guideline for the model. After the overall structure is completed and the individual components of the model have been assigned to the joints, inverse kinematics is applied to the limbs and head. In this manner, only one joint is positioned manually, and all the others that are further up in the joint hierarchy are automatically calculated by the software. Cited below are two videos of animals and three skeleton animations.

Demo videos
http://tethys.uni-ak.ac.at/cross-science/catwalk.mp4
http://tethys.uni-ak.ac.at/cross-science/jumping-kitten.mp4
http://tethys.uni-ak.ac.at/cross-science/jumping-leopard.mp4
dee http://tethys.uni-ak.ac.at/cross-science/leopard-jumping-in-real.mp4
http://tethys.uni-ak.ac.at/cross-science/floor-gymnastics.mp4

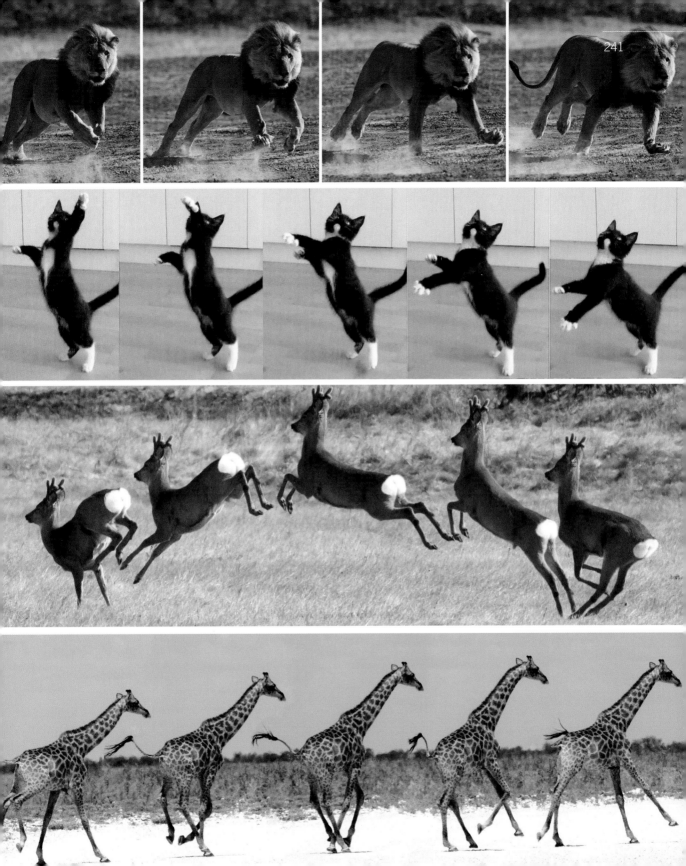

Index

Left: The Galton board simulates statistical distributions. In this application, it can be manipulated by making the falling balls bounce more or tend to fall to one side.

Demo video
http://tethys.uni-ak.ac.at/cross-science/galton-board.mp4

Right: With the help of statistical distributions as with the Galton board (left page), plants (shrubs, trees) can be created realistically. Certain parameters such as thickness, distance between branches and branching angle are pre-defined.

Demo video

http://tethys.uni-ak.ac.at/cross-science/tree-generator.mp4

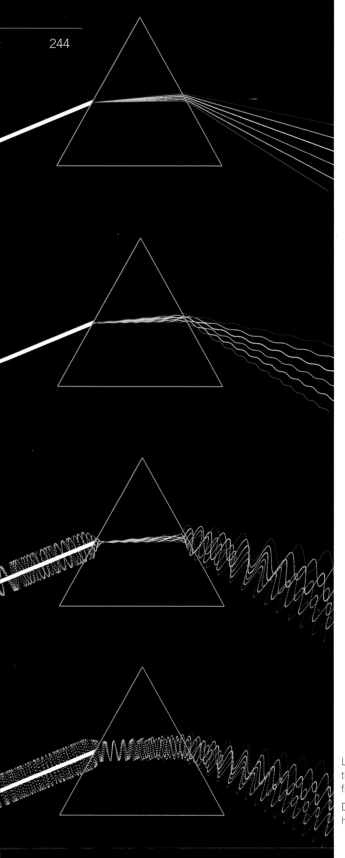

Left: An optical prism. The incoming light beam is fanned out twice into the colours of the rainbow. The matter can be understood both with refraction and the wave model of light.

Demo video
http://tethys.uni-ak.ac.at/cross-science/optical-prism.mp4

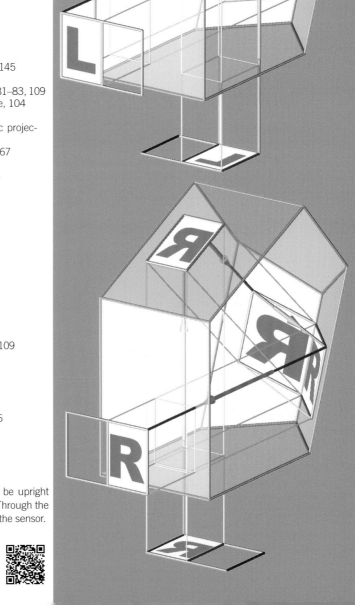

Right: An upside-down image generated by a lens system is to be upright via a mirror system and displayed in a viewfinder (SLR camera): Through the viewfinder, one should see exactly the image which then exposes the sensor.

Demo video
http://tethys.uni-ak.ac.at/cross-science/pentaprism.mp4

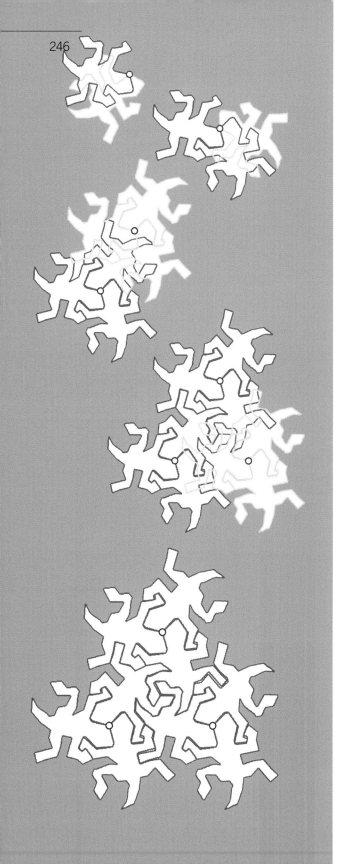

Left: The famous parquet by M. C. Escher with the lizards. The demo video shows how to create it through rotations and translations.

Demo video
http://tethys.uni-ak.ac.at/cross-science/galton-board.mp4

Right: Optimisation problem when filling a hemispherical bowl with balls of different sizes. The algorithm searches for the optimal packing density layer by layer.

Demo video
tethys.uni-ak.ac.at/cross-science/sphere-packing.mp4

Acknowledgements and Homepage

Co-author **Franz Gruber** passed away in 2019. Georg Glaeser has taken on the task of summarising the many joint works produced up to that point and to make them available to posterity. Additionally, topics were added that were in line with Gruber's areas of interest. The book is dedicated to his memory.

All photos are by Georg Glaeser.

Most of the 300 videos listed are from the authors.

The following videos are from other people:

Meda Retegan:
p. 11 (2nd video), p. 16 (1st video), p. 17, p. 36 (3rd and 4th video), p. 52, p. 63 (1st and 2nd video), p. 64 (1st, 2nd and 4th video), p. 65, p. 67 (2nd video), p. 212 (1st and 2nd video), p. 214, p. 216 (1st video), p. 217, p. 219 (2nd video), p. 248

Christian Clemenz:
p. 46, p. 76 (2nd video), p. 227, p. 239 (2nd video), p. 240 (1st, 3rd and 5th video), p. 90 (together with **Leonard Weydemann**)

Showcase TU Vienna (Geometry): p. 34 (3rd and 4th video)

Hans-Peter Schröcker: p. 26 (3rd video), p. 27 (2nd video), p. 227 (2nd video)

Simonas Sutkus: p. 76 (1st video), p. 94 (2nd video)

Nina Gstaltner: p. 151 (2nd video)

Lukas Kotolek-Steiner: p. 221 (2nd and 3rd video)

Claudia Carozzi, Stefan Felkel, and Christoph Fessl: p. 239 (1st video)

The author owes special thanks to **Eugenie Theuer** for the translation of the German version and **Julia Weber** and **Irene Karrer** for their conscientious proofreading.

The accompanying website

Please visit the website below. There you will find executable programmes that you can install (under Windows), as well as other demo videos that are not listed in the book, but are intended to stimulate your imagination. The advantage of this website is that it can be updated again and again after the book is printed.

Accompanying website
https://tethys.uni-ak.ac.at/cross-science/goodies/

Demo video
http://tethys.uni-ak.ac.at/cross-science/dragonfly-simulation.mp4